Children's Measurement:
A Longitudinal Study of Children's Knowledge and Learning of Length, Area, and Volume

Edited by

Jeffrey E. Barrett
Illinois State University
Normal, Illinois

Douglas H. Clements
University of Denver
Denver, Colorado

Julie Sarama
University of Denver
Denver, Colorado

JRME Monograph Number 16

NCTM | NATIONAL COUNCIL OF TEACHERS OF MATHEMATICS

1906 Association Drive, Reston, VA 20191-1502
(703) 620-9840; (800) 235-7566; www.nctm.org

Library of Congress Cataloging-in-Publication Data

Names: Barrett, Jeffrey Edward, editor. | Clements, Douglas H., editor. | Sarama, Julie, editor.
Title: Children's measurement : a longitudinal study of children's knowledge and learning of length, area, and volume / edited by Jeffrey E. Barrett, Illinois State University, Douglas H. Clements, University of Denver, and Julie Sarama, University of Denver.
Description: Reston, VA : The National Council of Teachers of Mathematics, [2016] | Series: JRME monograph series ; no. 16 | Includes bibliographical references.
Identifiers: LCCN 2016021515 | ISBN 9780873539890 (pbk.)
Subjects: LCSH: Geometry--Study and teaching. | Mathematical ability in children--Testing. | Educational surveys.
Classification: LCC QA461 .C4825 2016 | DDC 372.7/6--dc23 LC record available at https://lccn.loc.gov/2016021515

The National Council of Teachers of Mathematics is the public voice of mathematics education, supporting teachers to ensure equitable mathematics learning of the highest quality for each and every student through vision, leadership, professional development, and research.

Printed in the United States of America

Contents

Acknowledgments

We, the editors of this monograph as well as the principal and coprincipal investigators of the Children's Measurement Project, wish to acknowledge the participation of the prekindergarten to Grade 2 classroom teachers and school administrators at the Northeast site and the Grades 2–5 classroom teachers and school administrators at the Midwest site. We would also like to acknowledge the support of the graduate research assistants who helped with this research: Jason Ball, Katharine A. Bartelo, Kristen M. Bugos, Craig J. Cullen, Cheryl L. Eames, Nosisi Feza, Melike Kara, Jenni K. McCool, Jennifer S. McDonel, Amanda L. Miller, Lisa Napora, Katherine Newburgh, Chepina Rumsey, Jennifer A. Schiller, Douglas W. Van Dine, and Maria M. Vukovich.

The research reported here was supported in part by the National Science Foundation Research Grant (DRL-0732217). Any opinions, findings, and conclusions or recommendations expressed in this material are those of the authors and do not necessarily reflect the views of the National Science Foundation.

Introduction

This monograph is a report on a 4-year-long multisite longitudinal study in which we studied children's thinking and learning about geometric measurement (i.e., length, area, and volume). The Children's Measurement research team, as funded by the National Science Foundation between 2007 and 2012, completed the research reported in this monograph. During that time, our work had two emphases. The first and major emphasis was the study of the developing knowledge of continuous quantity in geometric measurement contexts of children from prekindergarten (pre-K) to Grade 5. The second emphasis was a retrospective accounting of the research methods and processes that we used to clarify and revise learning trajectories (LTs). Members of our research team at the Northeast site worked with the same group of children from pre-K to Grade 2, and members of our research team at the Midwest site worked with another group of children from Grade 2 through Grade 5.

ORGANIZATION OF THE MONOGRAPH

We present our findings with an introductory section and four subsequent sections: (a) background information about our project, including subsections on theoretical perspective, design, and participants in Chapter 1; (b) length in Chapter 2; (c) area in Chapters 3 through 6; (d) volume in Chapters 7 through 10; and (e) reflections on the work as a whole in Chapter 11. Our organization for the monograph follows a geometric progression from length to area to volume. We systematically present research findings to support our primary research questions addressing these three aspects of spatial measurement. We have already published many of the results on length in other venues. These findings, however, are foundational to all the other sections; therefore, we present a summary of our results to support our assumption that a generalizable LT is achievable, along with new findings and perspectives. We then introduce area and volume and present the findings for the early childhood and primary grades in one chapter, as well as the findings for the middle and later elementary grades in another. In the last chapter in each of these sections, we synthesize the findings to present implications regarding a revised LT, first for area and then for volume measurement.

Both the area and volume sections substantiated the usefulness of the LT construct and in addition revealed crucial extensions and refinements of the LTs. Specifically, we clarify how children learn and apply measurement knowledge in the early years (pre-K through Grade 5), in addition to exploring the development of children's cognitive abilities (e.g., spatial thinking, proportional reasoning) in understanding and using measurement strategies. In the monograph's final section (Chapter 11), we reflect on the work as a whole; present perspectives and conjectures on theoretical and methodological issues surrounding our efforts to conduct longitudinal research on LTs; and comment on the relationships among length, area, and volume measurement LTs.

Chapter 1

Measurement in Early and Elementary Education

Douglas H. Clements
University of Denver
Jeffrey E. Barrett
Illinois State University
Julie Sarama
University of Denver

THE IMPORTANCE OF MEASUREMENT

Quantitative reasoning and measurement competencies support the development of mathematical and scientific thinking from prekindergarten (pre-K) through Grade 8 (Clements, 2003; Davydov, 1991; So, 2013; Steffe, 1991; van den Heuvel-Panhuizen & Buys, 2008) and are fundamental to science, technology, engineering, and mathematics (STEM) education. Children's knowledge of quantity and strategies for quantifying, as well as their reasoning about quantity, are necessary elements of a working theoretical foundation for measurement (Case & Okamoto, 1996; Piaget, Inhelder, & Szeminska, 1960; J. P. Smith & Thompson, 2007; Steffe & Cobb, 1988; Steffe & Olive, 2010; Tal, 2013). In particular, geometric measurement (e.g., length, area, and volume) bridges the mathematical domains of number and geometry (Sarama & Clements, 2009b). Children forge this critical connection while they establish and reason about quantity throughout the mathematics curriculum.

Measuring is a nontrivial aspect of children's developing mathematical and scientific thinking. Its basis is in the exercise of comparative judgment; children compare magnitudes immediately or observe patterns of change over time by quantifying their experiences and observations. Measurement knowledge and strategies play broadly and deeply into children's understanding of both science and mathematics, making measurement a vital component of pre-K through Grade 8 curricula (So, 2013). For example, measurement experience provides images of variables as quantities in algebra, an important focus of the upper elementary and middle school curriculum, which does not often emphasize a conceptual approach to thinking quantitatively (A. G. Thompson & Thompson, 1996; P. W. Thompson & Thompson, 1994).

The *National Science Education Standards* (National Research Council, 1996) indicated that mathematical measurement plays an essential role in all aspects of scientific inquiry. C. L. Smith, Wiser, Anderson, and Krajcik (2006) pointed out the integrative role of measurement for learning science conceptually: "Given the centrality of measurement in science and the ways measurement can contribute to conceptual understandings, it is important to start early in developing a rich understanding of the measurement of important physical quantities" (p. 33). Measurement is a central aspect of spatial thinking; a quantitative understanding of space is often essential for posing questions, developing explanations, and communicating results (National Research Council, 2006; Presmeg & Barrett, 2003). Within technological design for spatial problems, children need to choose suitable tools and techniques and work with appropriate measurement methods to ensure precision and accuracy. The authors of the *Atlas of Science Literacy* (American Association for the Advancement of Science, 2001) described measurement concepts as a critical foundation for establishing evidence as well as a basis for modeling processes and systems (cf. Lehrer & Schauble, 2002, 2004; Petrosino, Lehrer, & Schauble, 2003).

In 2006, the National Council of Teachers of Mathematics (NCTM) released *Curriculum Focal Points for Prekindergarten Through Grade 8 Mathematics: A Quest for Coherence* (CFP). With this document, NCTM recommended that teachers focus primarily on target curriculum topics for each grade level from pre-K through Grade 8. The CFP offers guidelines for age-appropriate learning activities, clustered about central themes in mathematics. Furthermore, it portrays measurement as an essential topic, underscoring the important role that measurement instruction plays, especially before middle school. Although the National Governors Association Center for Best Practices (NGA) and the Council of Chief State School Officers (CCSSO) completed the Common Core State Standards for Mathematics (CCSSM; NGA & CCSSO, 2010) after we conducted the research reported in this monograph, the CCSSM also includes geometric measurement as an essential topic. In response to the fundamental need for developmental accounts based in research, we set out to furnish longitudinal accounts that would establish sequenced assessment goals across grades, guide curricular development programs, and strengthen teacher education.

THEORETICAL PERSPECTIVE

We approached this project through a theoretical lens called *hierarchic interactionalism* (Sarama & Clements, 2009b), a synthesis of previous theoretical frameworks that owes much to theorists and researchers too numerous to list here (see Sarama & Clements, 2009b) but that relies heavily on several (e.g., Carpenter, Franke, Jacobs, Fennema, & Empson, 1998; Confrey & Kazak, 2006; Karmiloff-Smith, 1992; Minsky, 1986; Piaget, 1941/1952; Siegler & Booth, 2004; Steffe & Cobb, 1988; Van de Rijt & Van Luit, 1999; Vygotsky, 1934/1986). This research corpus is the basis for the 12 tenets of hierarchic interactionalism (adapted from Sarama & Clements, 2009b):

1. *Developmental progression.* Children acquire content knowledge along developmental progressions of levels of thinking that are topic-specific (see Tenet 2). These progressions are consistent with children's intuitive background knowledge and patterns of thinking and learning, which may be culturally specific, albeit still influenced by "initial bootstraps" (see Tenet 6). Specific concepts and processes characterize each level of thinking.
2. *Domain-specific progression.* Developmental progressions are characterized within a specific mathematical domain or topic, such as measurement, and even measurement of a particular attribute.
3. *Hierarchic development.* Levels of thinking are internally coherent, but the learning process is often incremental and gradually integrative, rather than "stagelike."
4. *Cyclic concretization.* Developmental progressions proceed from sensory-concrete levels, in which perceptual concrete supports are necessary and reasoning is usually restricted to a few cases, to more explicit verbally based generalizations, resulting in integrated-concrete understandings relying on mental representations that serve as models for operations and abstractions (see Sarama & Clements, 2009a).
5. *Mutual development of concepts and skills.* Concepts and skills develop together, each supporting the further development of the other.
6. *Initial bootstraps.* Children are endowed with premathematical competencies and predispositions either at birth or soon thereafter; these competencies and predispositions support and constrain future learning.
7. *Different developmental courses.* Different developmental courses are possible; individual, environmental, and social confluences influence them.
8. *Progressive hierarchization.* Children integrate concepts and skills, building understandings that are hierarchical and that include both generalizations and differentiations.
9. *Environment and culture.* Environment and culture affect the pace and direction of development.
10. *Consistency of progressions and instruction.* Basing instruction on learning research adds value.
11. *Learning trajectories.* Curricula and instruction developed on the basis of full hypothetical learning trajectories are more effective than those that lack such a basis.
12. *Hypothetical learning trajectories.* Teachers interpret hypothetical learning trajectories and realize them through social interaction with children around instructional tasks.

Learning Trajectories

We based our longitudinal study of children's thinking and learning on the construct of a *learning trajectory* (LT). In his seminal work, Simon (1995) stated that a *hypothetical learning trajectory* (HLT) included "the learning goal, the learning activities, and the thinking and learning in which students might engage" (p. 133). We use the term learning trajectory rather than hypothetical learning trajectory for the sake of simplicity in our reference to LTs. Nevertheless, we note that the LTs are theoretical objects of research; as such, they are consistent with HLTs. We view LTs as

> descriptions of children's thinking and learning in a specific mathematical domain and a related, conjectured route through a set of instructional tasks designed to engender those mental processes or actions hypothesized to move children through a developmental progression of levels of thinking, created with the intent of supporting children's achievement of specific goals in that mathematical domain. (Clements & Sarama, 2004, p. 83)

Developmental progressions are the core component of LTs (and the main focus of this monograph), but they have two additional parts: goals and instructional activities. The synthesis of these three parts can be expressed as follows: to reach a certain mathematical competence in a given topic (the goal), children progress through sequential levels of thinking (the developmental progression), supported and assisted by tasks (the instructional activities; see Clements & Sarama, 2004).

Mathematical goal. We determined our goals through a synthesis of three sources. First were the big ideas and core competencies of the domain—clusters of concepts and skills that are mathematically central and coherent, consistent with children's thinking, and generative of future learning. These big ideas and competencies come from several large projects, including those from the NCTM's (2006) CFP, the National Mathematics Advisory Panel's (2008) final report, the CCSSM (NGA & CCSSO, 2010), and related projects (Clements, Sarama, & DiBiase, 2004). For example, at least eight concepts form the foundation of children's understanding of length measurement: understanding of the attribute, conservation (a good example of a concept often assumed by content descriptions, but important psychologically), transitivity, equal partitioning, iteration of a standard unit, accumulation of distance, origin, and relation to number (Clements & Stephan, 2004; Sarama & Clements, 2009b). Our goal for children certainly includes that they learn to "measure and estimate lengths in standard units" (NGA & CCSSO, 2010, p. 20), including selecting appropriate tools, relating measurements by using different units, estimating, and comparing measurements. In addition, we include children's explicit understanding of how such competencies relate to the concepts of equal partitioning, iteration, and origin, as well as others. As an example in measurement of two or three dimensions, the cognitive competence of spatial structuring is fundamental but ignored in most content-only descriptions of mathematical goals.

Developmental progression. A main assumption is that children acquire most content knowledge, such as knowledge of geometric measurement, along developmental progressions of levels of thinking that are particularly consistent with children's intuitive knowledge and patterns of thinking and learning (at least in a particular culture, but guided in all cultures by innate competencies). That is, children progress through domain-specific levels of understanding in ways we can characterize by the concepts and processes (mental objects and actions on them) that build hierarchically on previous levels. These actions on objects are children's main way of operating on, knowing, and learning about the world, including the world of mathematics. As a brief example, our previous work on length LTs revealed that children reliably developed through the levels of thinking, passing through a level in which they measured length by placing multiple units, sometimes leaving gaps between units, before they could iterate a unit and then measure length competently with multiple strategies and tools (Barrett et al., 2011; Sarama, Clements, Barrett, Van Dine, & McDonel, 2011).

Developmental progressions follow a pattern. Initially, at the sensory-concrete level, children need perceptual concrete supports. Next, children create verbally encoded generalizations that have distance from those supports. Finally, children construct integrated-concrete understandings on the basis of mental representations that serve as models for mathematical concepts and operations (Clements & Sarama, 2009; Sarama & Clements, 2009a).

Levels of thinking are coherent, with increased sophistication, complexity, abstraction, and generality often characterizing higher levels (Clements & Sarama, 2014b; Confrey, Rupp, Maloney, & Nguyen, 2012; Maloney & Confrey, 2010; Sztajn, Confrey, Wilson, & Edgington, 2012). Conceptualizing levels at different grain sizes is possible. As an example, the "growth points" of the Early Numeracy Research Project in Australia are developmentally widely spaced (e.g., children might achieve successive points in different grades; Clarke et al., 2002). Our approach (Sarama & Clements, 2009b) is to find levels of finer grain size so that, for example, teachers can see and observe development through several levels within their grade but still retain the defining characteristics of levels, such as constancy (some properties, states, or activities remain consistent at the level), order invariance, and hierarchic incorporation and integration (Sarama & Clements, 2009b; Steffe & Cobb, 1988).

Instructional tasks. Although the emphasis of this research project was the developmental progressions, instructional tasks, the third component of LTs, also played a role. In hierarchic interactionalism, educators often design tasks to present a problem that is just beyond the children's present level of operating. In some cases, sequences of tasks may furnish sufficient goal-directed activity, changing the focus of attention (e.g., from a physically concrete model to a mental image or symbolic representation), reflection (identification and connection of commonalities), and anticipation (prediction of the result of an activity before carrying it out) to engender children's generalization to form abstractions, such as creating a new object (e.g., a unit of units) that is more general or abstract mathematically (Simon, 2013). In other cases, children may need to actively engage in reformulating the problem or their solution strategies, often with peer interaction and teacher guidance. In reflecting on their activity, children learn whether they have solved the original problem or whether they need to engage in more thinking. This cycle may continue until they build a new level of thinking. A critical mass of such construction in individuals and small groups often allows productive class discussions that can justify, formalize, and symbolize the mathematics (Simon, 2013) and eventually integrate it into a mathematical system (van Hiele, 1986; van Hiele-Geldof, 1984) in a manner appropriate for the children.

Our theory of hierarchic interactionalism does not attempt to include or categorize a full range of teaching and learning processes (e.g., see Clements, Agodini, & Harris, 2013; Clements & Sarama, 2009; Tharp & Gallimore, 1988; van Hiele-Geldof, 1984) but posits that instructional tasks are a main component of effective instruction and that any pedagogical practice is effective to the extent that it activates children's mental actions on objects that support the subsequent level of thinking in the development progression.

Misconceptions Regarding Learning Trajectories

Because of the complexity of LTs, it is no surprise that educators often misunderstand them. One claim is that they are nothing new (e.g., are just "scope and sequences"). The LTs of hierarchic interactionalism share characteristics with previous research programs but differ in significant ways (for a complete analysis, see Clements & Sarama, 2014b). LTs build on previous work, which has developed from simple accumulations of connections (Thorndike, 1922) to more complex views of thinking and learning (Gagné, 1965/1970; Piaget et al., 1960; Resnick & Ford, 1981). All theories attempt to explain psychological sequences, with previous accounts relying on the acquisition of facts and skills over time. LTs include such sequences but are not limited to sequences of "pieces" of knowledge only. They contain descriptions of children's levels of thinking; and it is therefore impossible to summarize them by stating a mathematical definition, concept, or rule (cf. Gagné, 1965/1970). Levels of thinking convey *how* children think about a topic and *why*. Finally, earlier theories often based instruction on a transmission approach. LTs have an interactionalist view of teaching—children interacting with a teacher and other children around instructional tasks, with the teacher using all three components of LTs to guide those interactions.

A second misconception is that LT levels rigidly categorize children (i.e., "put children in boxes"). In contrast, the theory posits that children can operate at multiple levels, with most children working mostly at one level or in transition between two levels. We identify children as being at a certain level when most of their behaviors reflect the patterns of thinking at that level. Often, they show a few behaviors from the next (and previous) levels while they learn. The continued existence of earlier levels, as well as the role of intentionality and social influences in their instantiation, explains why in some contexts even adults fall back to earlier levels (e.g., failing to conserve in certain situations).

Related to the second misconception is a third: the belief that LTs are insensitive to individual and cultural differences. Hierarchic interactionalism posits that different developmental courses are possible within constraints, depending on individual, environmental, and social–cultural confluences (Clements & Battista, 2001; Confrey & Kazak, 2006; Sarama & Clements, 2009b). Within any developmental course, children at each level of development have a variety of cognitive tools—concepts, strategies, skills, utilization, and situation knowledge—that coexist. The differences within and across individuals create variation that is the wellspring of invention and development. At a group level, however, these variations are not wide enough to vitiate the theoretical or practical usefulness of the tenet of developmental progressions; for example, in a class of 30, one might find only a handful of different solution strategies (cf. Murata & Fuson, 2006), most of which represent adjacent levels along the developmental progression. Further, environment and culture affect the pace and direction of the developmental courses. For example, the degree of experience that children have to observe and use number and other mathematical notions and to compare these uses affects the rate and depth of their learning along the developmental progressions. The degree to which children learn mathematical words, exposure to which varies greatly across cultural groups (Buss & Spencer, 2014), affects developmental courses. Words alert children to the class of related words that they must learn and to specific mathematical properties, laying the foundation for learning mathematical concepts and language (cf. Sandhofer & Smith, 1999) by providing a nexus on which to build their nascent constructs (Vygotsky, 1934/1986).

On one hand, no single absolute developmental progression, and thus LT, exists for a mathematical domain because other factors—such as environment, culture, and educational experiences—may affect a child's actual pathway for learning and development. On the other hand, a large number of pathways do not exist, because universal developmental factors interact with culture and mathematical content. Educational innovations and culturally specific educational environments may establish new and potentially more advantageous sequences, so work with LTs should recognize the culturally situated nature of children's learning and the need for a "critical view" (Wager & Carpenter, 2012). We concur that educators should examine generalized claims about children by including adequate samples of a broadly inclusive collection of children across cultural and economic constructs. Yet, this respect for varying populations of children need not imply that we as an educational community must expect that substantially different LTs exist for some groups of children. First, people often translate this expectation into "my children learn more slowly," an avoidable trap of low expectations disguised as sensitivity to cultural and individual differences. Second, we believe that the cognitive core of the LTs is valid within different cultural contexts and should be instantiated in different ways in different sociocultural settings to promote learning and equity. Of course, researchers should test this belief in future research that uses such a cognitive core account in a range of settings.

In summary, we used LTs in our research because they address the components of mathematical goals, developmental progressions, and instructional tasks; and they are increasingly important in guiding the writing of standards (Maloney, Confrey, & Nguyen, 2014; NGA & CCSSO, 2010) and curricula (Clements, 2007; Maloney et al., 2014). Further, teachers who understand LTs understand the mathematics associated with the various levels of reasoning, the way that children think and learn about mathematics, and how to help children learn it better. LTs connect research and practice. They help teachers understand the level of knowledge and thinking of their classes and the individuals in their classes as important in serving the needs of all children. In this way, LTs serve as a basis for formative assessment, an essential component of high-quality teaching (see National Mathematics Advisory Panel, 2008). More refined and better validated LTs can contribute to improved standards, curricula, and teaching practices.

THE NATIONAL SCIENCE FOUNDATION'S CHILDREN'S MEASUREMENT PROJECT

The Children's Measurement research team built, refined, and tested a set of LTs for geometric measurement (Barrett et al., 2011; Barrett et al., 2012; Sarama, Clements, Barrett, et al., 2011; Sarama, Clements, Van Dine, et al., 2011) to represent children's developing measurement knowledge and their ways of learning measurement. To complement our emphasis on children's learning, we looked for ways of improving instruction, assessment, and curriculum development related to measurement concepts. A main goal of our team has been to address the educational need for measurement in the pre-K to Grade 5 mathematics curriculum by clarifying, extending, and improving LTs. We see LTs as tools with wide-ranging application in education. Because of our interest in addressing a range of educational issues, we work to improve these tools through design cycles of research. For example, after establishing a mathematical goal (e.g., NCTM, 2006; NGA & CCSSO, 2010), we began by reviewing the research and attempting to piece together a coherent longitudinal story, even from separate disparate studies (Sarama & Clements, 2009b). We assessed children at the beginning and end of the study by using clinical interviews with younger children and a written format with older children to clarify the cognitive attributes of the nascent levels. We conducted teaching experiments with individual children, as well as classroom-based teaching experiments with intensive qualitative analyses to revise and expatiate those levels and their interconnections and sequences.

The research in this monograph began with the LTs that we had created by using these procedures (Clements & Sarama, 2014a; Sarama & Clements, 2009b), recognizing that we needed to refine and extend them. We believe that the resulting LTs can strengthen the educational and curricular infrastructure needed for implementation of the measurement standards within the CCSSM (NCTM, 2006; NGA & CCSSO, 2010) from kindergarten through Grade 8.

Our Research Goals

One of our central goals has been to delineate the development of children's knowledge and potential strategic competence on increasingly demanding sets of length, area, and volume measurement tasks through successive grade levels from pre-K to Grade 5. Another central goal has been to establish prototypical longitudinal stories of children's ways of reasoning through measurement activities. Such stories, which we based on increasingly complex tasks, influenced the design and implementation of instructional sequences, guiding teachers toward essential concepts relevant to children in their particular grade levels, in keeping with the CFP (NCTM, 2006). We therefore conducted longitudinal studies by following the developmental reasoning of selected children at two research sites over a 4-year span, from pre-K to Grade 2 at one research site and from Grade 2 to Grade 5 at the other (see Participants section for more detail).

We expect these LTs for measurement to help improve formative classroom assessment by teachers because they can more easily recognize opportunities to engage children proactively in conceptual change (Sztajn et al., 2012; Wickstrom, Baek, Barrett, Cullen, & Tobias, 2012; Wilson, 2009). We also expect the LTs to support formal assessment efforts and research on curriculum development (Clements, 2007).

Research Question

Our research focuses on supporting developmental coherence within the curriculum for elementary mathematics and within the instructional approaches that educators recommend for teaching. We are therefore interested in describing the growth of children as persons who reason about their world by using measurement tools to solve problems and organize quantitative thinking (K. F. Miller, 1989; P. W. Thompson, 1992). In particular, we examine central concepts in measurement learning, including the following: (a) identifying attributes of continuous dimensions to quantify and compare, (b) building appropriate units with conservation, (c) partitioning objects to generate quantity, (d) composing and coordinating units and groups of units in structures cumulatively, (e) iterating units in correspondence with number schemes including an assignment for zero, and (f) marshaling logical support and justification of measurement claims to guide scientific reasoning (cf. Clements & Stephan, 2004; Lehrer, 2003; Stephan & Clements, 2003). The main research question that motivated the Children's Measurement Project addresses these concepts across three spatial measurement domains (length, area, and volume): How do children's thinking and learning about each of the three domains of spatial measurement develop over time from pre-K through Grade 5?

Participants

We drew the children in our study from an urban parochial school in the Northeast (78% White, 9% Black, 9% Hispanic, 1% Asian, 3% two or more races) and from a suburban public school in the Midwest (68% White, 9% Black, 9% Hispanic, 9% Asian, 6% two or more races). Approximately 20 children were in each class at these schools, and all the children in pre-K through Grade 5 at both schools participated in a quantitative assessment conducted during Year 1 of the study. By using this assessment and collaborating with teachers, we selected a cohort of eight pre-K children at the Northeast site and eight Grade 2 children at the Midwest site. They became the focus children for our 4-year longitudinal study. Children at the Northeast site participated in the study during their 4 years of schooling from pre-K through Grade 2. Children at the Midwest site participated in the study during their 4 years of schooling from Grade 2 through Grade 5.

On the basis of our initial quantitative assessment and teacher input, we selected two low-performing children, four middle-performing children, and two high-performing children as a representative subset at both school sites. In addition, we strove for variation in socioeconomic status (SES) and cultural background that reflected the demographics of the schools. The selected sample also included equal numbers of male and female children. Finally, we used the same methods to select another eight children at each site to serve as "background" children. We used these background children to pilot tasks, gather additional information, and act as replacements for children who left the focus group. Attrition at the Northeast site resulted in three children leaving the study after the first year, one child after the second year, and two children after the third year. At the Midwest site, one child left the study after the first year, one child after the third year, and two children during the fourth year. Each time a child left the study, a child from the background group who had comparable characteristics replaced the child who left so that we could maintain a balanced focus group (in gender, cultural background, mathematics performance, and SES) at that site.

Project Design

The next two sections describe our qualitative and quantitative methods for data collection and analysis. Although our primary goal was to provide longitudinal accounts of children's developing understanding of measurement, we complemented these accounts by designing and checking assessment items at each level of the length, area, and volume LTs. We administered these assessment items to all the children at the Midwest site and selected classrooms of children at the Northeast site.

To produce longitudinal accounts of children's developing knowledge, we used teaching experiments to create descriptions of plausible sequences of growth along the LTs for length, area, and volume. These teaching experiments functioned in a supportive role alongside regular classroom instruction based on existing curriculum materials in the schools. We recognize that a number of instructional factors may have led to the documented growth through extended sequences of teaching episodes throughout the study. Our primary goal, however, was to validate and refine the developmental progressions within the LTs instead of focusing on the instructional tasks.

Qualitative Methods

At the Northeast site, three children remained in the study for the 4-year duration, and they were the main focus of our qualitative analysis:

- Edith was a high-performing female with a desire to provide the correct answer always. She sought approval and positive feedback from the researchers, so that the researchers had to exercise vigilance to avoid guiding her through their vocalizations and behavior. Edith often commented that her mother regularly did mathematics problems with her at home.
- Ryan was a middle-performing male who was often quick to answer and usually took the lead in any group setting. He had a strong number sense and a willingness to take risks, which often resulted in repeated attempts to work through problems until they were complete.
- Lia was a low-performing female with a strong sense of spatial structuring: She was the first to correctly construct a three-dimensional cube object from a two-dimensional representation. However, she struggled to estimate and often lost track during long multistep problems. Lia remained in kindergarten for 2 years because of weak recognition of letters.

To furnish additional details, we turned to two children for whom we had between 2 and 3 years of longitudinal data: Zola, a middle-performing female and Marina, a low-performing female. Although the school retained Lia in kindergarten, the other four children progressed yearly through the grades.

At the Midwest site, five children remained in the study for the 4-year duration, and we focused on them for our qualitative analysis. We selected the children from two classes of children at one school.

- Arielle was a high-performing female with a tendency to use arithmetic operations to operate on problems posed in geometric space rather than operating with spatial operations and images directly to solve problems. She also had a tendency to work quietly until reporting her response to tasks.
- Abby was a middle-performing female with a tendency to explore task situations with a variety of strategies, and she often appealed to spatial actions or objects to support her reasoning about tasks. She was easy to engage and was willing to explain her thinking while she was working.

- Owen was a middle- to high-performing male who often worked quietly but with keen energy to examine problems in measurement. He balanced his reasoning between spatial and arithmetic operations.
- Drew was a middle-performing male with an articulate sense of mathematics, a capacity to imagine space and operations in space, and a willingness to use novel approaches to problems of measurement. However, sometimes his novel approaches took him off task and he required extra work to regain the task topic and goal.
- Anselm was a middle-performing male oriented primarily toward reasoning about spatial objects, with a secondary emphasis on arithmetic operations with measured values. We noticed that Anselm often focused on the instructions and questions of his peers or teacher to such an extent that he would prefer to follow another person's line of questions or thinking and develop alternative approaches to problems rather than proceed with his own initial approach.

To provide additional analysis at various stages of our study at the Midwest site, we included further data from two children for whom we had between 2 and 3 years of longitudinal data: Danny, a low-performing male, and Randy, a middle-performing male. Finally, we include occasional analysis from sessions with one female child, Sara, for whom we had little more than 1.5 years of data. Sara was a low-performing female who was unable to complete the second year of this research program when her family moved away from the school.

Our qualitative data sources included assessments, as well as individual and classroom-based teaching experiments. We administered assessments to children during 2008 and 2011. These two sets of assessments complemented the teaching experiments. The assessments included modifications of initial tasks and novel tasks designed to elicit specific thinking and behaviors indicative of particular levels of the developmental progression of the LTs (see www.childrensmeasurement.org).

To further evaluate each LT beyond the assessments, we conducted two types of teaching experiments (Cobb & Gravemeijer, 2008; Steffe & Thompson, 2000). First, we conducted individual teaching experiments, which included teaching episodes that occurred approximately every 4 weeks with individual focus children. Second, classroom-based teaching experiments occurred approximately once a year; during these experiments, we took the lead in the measurement instruction when it occurred in the classroom curriculum (see Figure 1.1). Previous observations and interpretations influenced each subsequent teaching episode. We designed the classroom-based teaching experiments with the classroom teachers. We also observed the classroom teachers' lessons on measurement. Data sources included video records and field notes. This combination of individual and classroom settings addressed our desire to study children's behaviors in naturalistic contexts. We analyzed these behaviors and compared them with the LTs to identify the level at which a child was demonstrating knowledge or to highlight gaps or inconsistencies in the LT description, and we tentatively revised LTs when a preponderance of evidence favored revision. The most demanding checks were across investigators within and especially between sites. We had to confirm any consistent behavior or revision across sites and incorporate or explain any disconfirming evidence before we could consider that the theoretical assertion was warranted.

	Northeast Site	**Midwest Site**
Year 1 2009–2010 school year	• Initial assessment • 4-day CTE focusing on length • 4 teaching episodes: 4 involving length and 1 also involving area	• Initial assessment • 4 teaching episodes: 3 involving length and 1 involving area
Year 2 2010–2011 school year	• 4-day CTE focusing on area and perimeter • 13 teaching episodes: 11 involving length, 4 involving area, and 7 involving volume	• 4 CTEs: 3 focusing on length and 1 focusing on area • 9 teaching episodes: 4 involving length, 2 involving area, and 3 involving volume
Year 3 2011–2012 school year	• 5-day CTE focusing on length • 14 teaching episodes: 8 involving length, 5 involving area, and 6 involving volume	• 5 CTEs focusing on area • 12 teaching episodes: 8 involving length, 7 involving area, and 6 involving volume • Grade 4 assessment
Year 4 2012–2013 school year	• 9 teaching episodes: 2 involving length, 6 involving area, and 4 involving volume • 8 Measurement Club meetings: 5 involving length, 5 involving area, and 6 involving volume • Final assessment	• 2 CTEs involving length, area, and volume • 9 teaching episodes: 8 involving length, 6 involving area, and 3 involving volume • Final assessment

Figure 1.1. Data collection cycles and data sources across the 4-year longitudinal study. (A CTE is a classroom-based teaching experiment.)

Quantitative Methods

To provide triangulation on the qualitative findings and to ascertain whether we could generalize those findings to other children, we created an assessment instrument and administered it to children (n = 258) in pre-K through Grade 5 at the two participating schools during spring 2011. These children were a representative sample of students at their respective schools. The total sample included 25 pre-K, 32 kindergarten, 29 Grade 1, 42 Grade 2, 41 Grade 3, 44 Grade 4, and 45 Grade 5 children. This sample included the cohorts of children that we followed in our teaching experiments.

We developed the instrument analyzed here to align with the levels of our measurement LTs, specifically those for length, area, and volume. Tasks included the measurement items from a previously developed and validated assessment, as well as tasks from previous empirical studies (Clements & Sarama, 2007). We used two items to assess each level within each of these LTs (for a total of 52 items; see Figures 1.2, 1.3, and 1.4 for sample length, area, and volume items, respectively), with the understanding that a correct response on an item demonstrated that a child was at least at that level. In addition, for the lower levels of each LT, we designed items for presentation through interview to limit the confounding effect of reading ability for younger children. We assessed children in pre-K, kindergarten, and Grade 1 entirely through interview; we assessed children in Grades 2 and 3 through a combination of interview and written items; and we assessed children in Grades 4 and 5 entirely through written items. We videotaped all interviews and conducted them one-on-one with an assessor and child. We presented all written items to children in their classrooms and collected all work and written responses.

Text continues on page 20

Length Direct Comparer

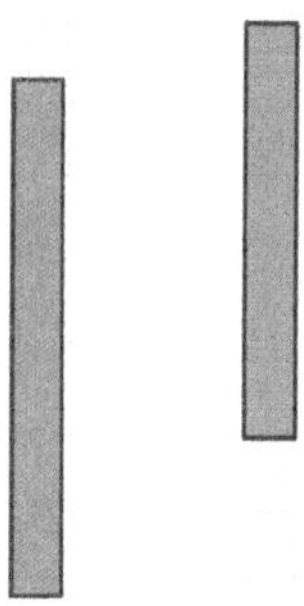

longer shorter

Place the two strips of paper (one longer and one shorter) on the desk in front of the child. Place them about 10 to 12 inches apart in the orientation shown in the illustration. Allow the child to move them.

Ask: *Which strip of paper is longer (or taller)? How do you know?*

End-to-End Length Measurer

Put out a 12-inch strip and a pad of 1.5-by-2-inch sticky notes. Place the pad of sticky notes on the tabletop at an angle of about 45 degrees with respect to the table edges.

Ask: *How many sticky notes will fit along this strip?*

Figure 1.2 continues on next page

Length Unit Relater and Repeater

Say: *I measured this long strip* (hold up 12-inch green strip) *with this* (hold up 1-inch yellow strip) *and found that it was 12 yellow strips long. If I measure the long strip with this blue strip* (hold up 2-inch blue strip), *how many of these blue strips will I need?*

Consistent Length Measurer

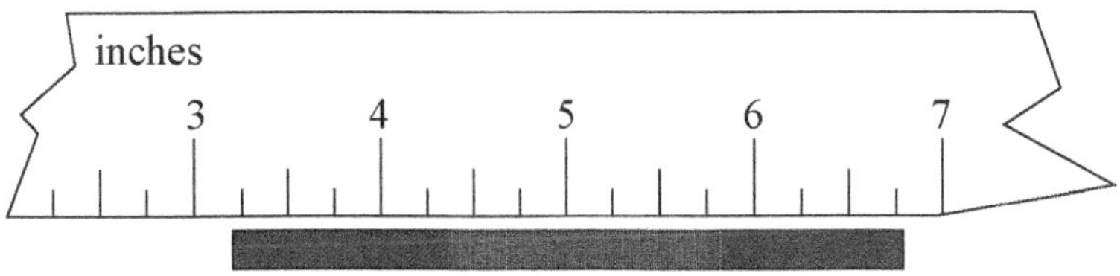

Say: *This is a picture of a rod just below a broken section of a ruler. Use this picture to measure the length of the rod. How long is the rod?*

Figure 1.2. Sample length items from the assessments.

Area Simple Comparer

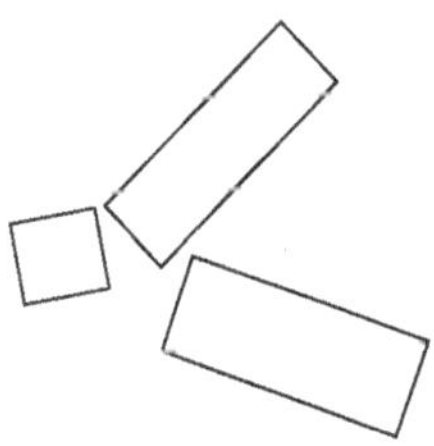

Scatter the rectangle cutouts in front of the child. (Make sure that they do not line up.) Allow the child to move the rectangles.

Ask: *Which piece of paper will let you paint the biggest picture?* (If the child wants to fold or cut, ask, *Can you do it without folding or cutting?)*

Side-to-Side Area Measurer

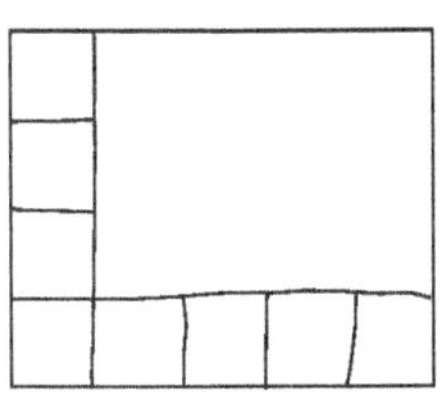

Say: *I wanted to cover this rectangle* (trace around the boundary of the larger rectangle) *with these squares* (point to one of the square-inch units). *I started drawing them in. Please finish the drawing by completely covering the rectangle.*

Figure 1.3 continues on next page

Area Unit Relater and Repeater

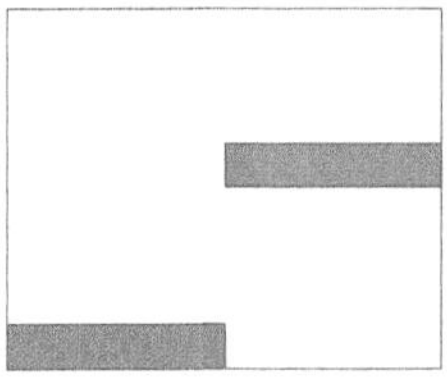

Ask: *How many tiles like this one*  *would cover the larger rectangle? Please include the two tiles already drawn.*

Partial Row Structurer

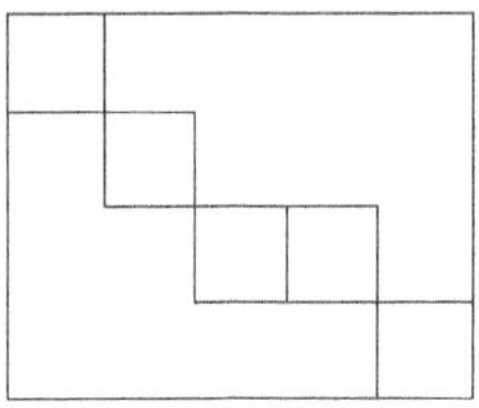

Ask: *How many squares would completely cover this rectangle? Please include the five squares already drawn.*

Area Row and Column Structurer)

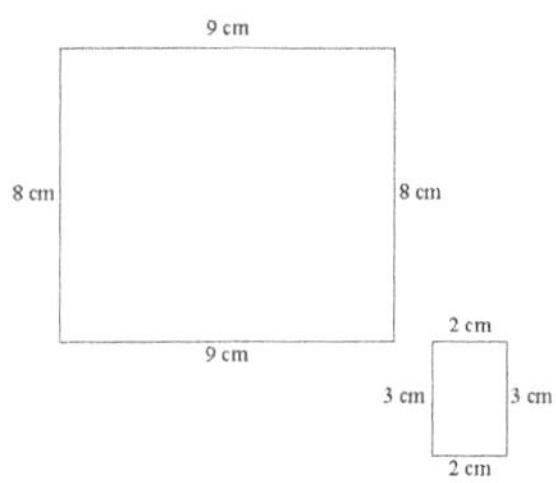

Ask: *How many of the small rectangles would cover the large rectangle?*

Figure 1.3. Sample area items from the assessments.

Capacity Direct Comparer

Show the child the two containers as shown below:

Say: *Pay attention because I am going to ask you a question about these two containers in a minute.*

(Point to the two containers. Completely fill one of the containers with water. Pour the water from container into the other.)

Ask: *Which of these two containers can hold more water?*

(Point to the two containers again.)

Primitive 3-D Array Counter

Place a 2 × 3 × 2 solid ("glued together") on the table. Show a separate inch cube.

Say: *This is a cube* (show child individual inch cube).

Ask: *How many cubes like this do you think that you would need to make this?* (gesture to solid)

Partial 3-D Structurer

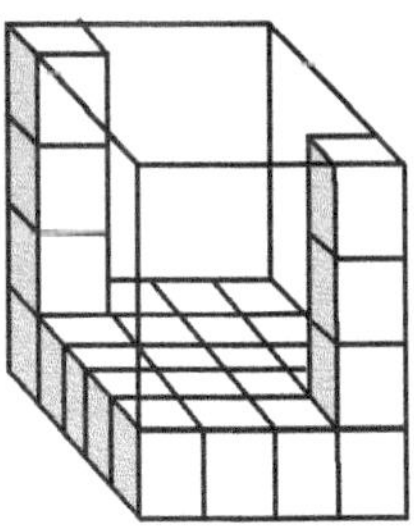

Ask: *How many cubes altogether will it take to fill the box?*

Figure 1.4 continues on next page

3-D Row and Column Structurer

Tell students: *This net* (on the left) *folds up to make the box* (in the middle).

Ask: *How many cubes* (on the right) *would you need to fill the box?*

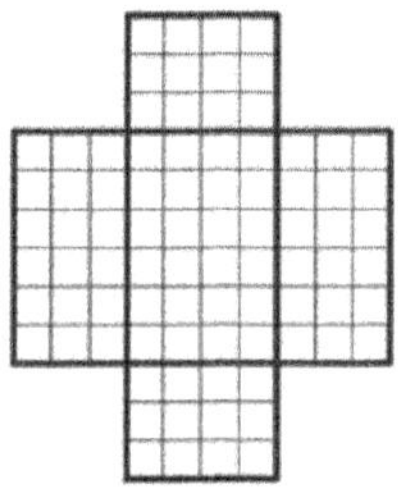

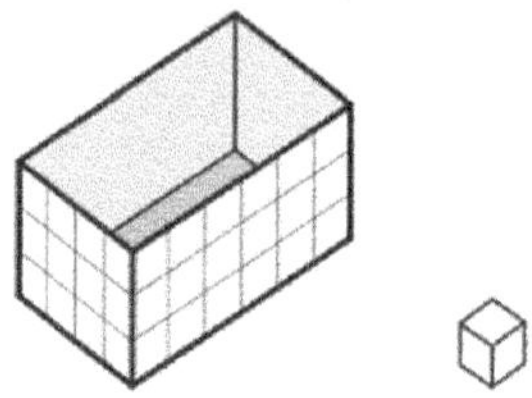

Figure 1.4. Sample volume items from the assessments.

We pooled these data across both the Northeast and Midwest sites and coded each item for correctness according to the following rubric: 0 = incorrect, 1 = correct with prompt (all prompts given according to the assessment protocol), 2 = fully correct. We considered any item coded with 1 as partially correct in the Rasch modeling. During the coding process, we included an additional code of 999 to identify any item for which we could not attribute a child's response directly to the child's understanding (e.g., assessor error affected the response of the child, child's response was unintelligible or unreadable, or child did not provide a response for the item). We treated any item coded as 999 as missing data in the analysis.

In our analysis of these data, we used item response theory (IRT), which allows researchers to create an interval scale of scores for both the difficulty of items and the ability of the persons assessed. The Rasch model is the simplest and most efficient IRT model. Use of the Rasch model furnishes evidence of both the validity and the reliability of an assessment. In addition, it allows mathematical estimation of both the probability that a person will answer an item correctly (person ability), as well as the probability that an item will be answered correctly by a person (item difficulty). Rasch modeling yields an ability score on an interval scale with a consistent, justifiable metric, thereby allowing accurate comparisons, even across ages, as well as meaningful comparison of change scores (Wright & Stone, 1979). Rasch modeling reports all ability scores in units called *logits*. Just as two inches is twice as long as one inch, two logits represent twice as much ability as one logit. This scoring differs from other scores such as percentile ranks, for example, because we cannot say that a person at the 50th percentile has twice as much ability as a person at the 25th percentile.

One underlying assumption of Rasch modeling is that we are measuring a unidimensional construct or latent trait. For our analysis, we defined *measurement competence* as that latent trait (Bond & Fox, 2007; Linacre, 2014; Watson, Callingham, & Kelly, 2007). To measure measurement competence, we sequenced the items, strictly maintaining the order within each measurement domain (length, area, volume) but intermingling items across domains according to the available developmental evidence, including age specifications from the literature and difficulty indices from our pilot testing. We therefore posited that items were organized according to increasing order of difficulty across domains, but our theoretical claims that this sequencing represented increasingly sophisticated levels of mathematical thinking were only for items within a given domain. We submitted the results of administering this revised instrument to the Rasch model.

Rasch analyses estimate the distance between items and between persons on a single scale. That is, item difficulty and each person's underlying competence are on the same equal-interval scale, indicating the theoretical latent trait. We used Winsteps (Linacre, 2014) to estimate fit statistics, reliabilities, separation indices, and item difficulties. Fit statistics (infit and outfit) are estimates of the degree to which responses show adherence to the expectations of the Rasch model. They indicate how well the model empirically supports the assumption of unidimensionality; that is, whether it is measuring a single attribute (a critical characteristic of fundamental measurement). In addition, the mean square (MNSQ) statistic is a transformation of the difference between the predicted and the observed scores (residuals) that indicates the degree of fit of an item or a person. Its expected value is 1, with values between 0.5 and 1.50 regarded as productive for measurement (Wright & Linacre, 1994). The Z-statistic (ZSTD) is a standardized fit statistic with a mean of 0 and variance of 1. For ZSTD, the range of acceptable values for a 95% confidence interval is between −2 and 2 (Bond & Fox, 2007; Linacre, 1994).

Item reliability is an estimate of the replicability of item placement within the hierarchy of items along the measured trait and is similar to Cronbach's alpha (Bond & Fox, 2007; Linacre, 1994). Item separation indices are assessments of the ability of the measure to differentiate items along the scale. Because we had employed qualitative analyses to refine the instrument in several previous cycles of formative testing and revision, we performed qualitative examination of video only on items with poor item characteristics for this study.

Chapters 2, 3, and 7 of this monograph give a description of specific Rasch analyses of the separate dimensions for length, area, and volume, respectively. These specific analyses provide further quantitative support in the validation and refinement of the developmental progressions in each measurement domain.

SECTION I

Chapter 2

Length: A Summary Report

Douglas H. Clements
University of Denver
Jeffrey E. Barrett
Illinois State University
Julie Sarama
University of Denver
Craig J. Cullen
Illinois State University
Douglas W. Van Dine
Metropolitan State University of Denver
Cheryl L. Eames
Southern Illinois University, Edwardsville
Melike Kara
Towson University
David Klanderman
Trinity Christian College
Maria Vukovich
The Center for Victims of Torture

Length and distance are measures of one-dimensional extent. The *length* of an object is the quantification of how far it is between the endpoints of an object. Similarly, *distance* is the quantification of how far it is between any two points in space. Measuring these characteristics involves identifying a unit of length and subdividing the object by that unit. Subdividing and unit iteration are complex mental accomplishments that measurement curriculum materials and instruction too often ignore. Therefore, many researchers go beyond the physical act of measuring to investigate children's understandings of measuring as covering one-dimensional space and quantifying that covering (e.g., Clements, 2003).

This chapter summarizes research relevant to building an improved developmental progression for a learning trajectory (LT) for length measurement. We begin with a brief review of theory and research on length measurement, including that leading to the development of our initial LT (Clements & Sarama, 2009; Sarama & Clements, 2009b). Then we describe the research that we conducted to test and revise the developmental progression of that initial LT. We describe the quantitative results for length from our final assessment and then—because most of the findings have been published elsewhere—we summarize the qualitative research results and their implications for revisions to particular levels of the initial developmental progression. We include charts that illustrate the growth of each child throughout the 4-year longitudinal study, as well as composite growth charts for the entire group of children from each of the two research sites, one spanning pre-K through Grade 2 and the other spanning Grade 2 through Grade 5. We conclude by presenting the revised levels of the new developmental progression component of the LT for length measurement with a rationale for the revisions.

REVIEW OF RESEARCH ON LENGTH MEASUREMENT

Piagetians defined the concept of length measurement as the cognitive synthesis of subdivision and change of position, which involved taking one part out of the whole and iterating that unit along the whole (Piaget, Inhelder, & Szeminska, 1960). Their experiments showed that young children appeared to believe that placing an object between two positions changed the distance between these positions (Miller & Baillargeon, 1990). However, subsequent research indicated that children 3.5 to 5 years of age seemed to understand these concepts, at least in some task contexts (Bartsch & Wellman, 1988).

Similarly, Piagetians argued that processes such as indirect comparison by using transitive reasoning were not possible for children until they acquired conservation of length (e.g., Kamii & Clark, 1997; Piaget et al., 1960). However, approximately 40% of 4-year-olds could avoid errors on direct and indirect routes and even gave correct explanations (Fabricius & Wellman, 1993). Further, altering the Piagetian conservation-of-length task did elicit conservation responses from young children (Schiff, 1983). That is, rather than the traditional task of determining whether two equal-length sticks remained equal in length after researchers moved one of them, researchers showed children a stick that fit into a box. After the researcher moved the stick, the researcher asked children whether the stick would still fit exactly into the box. Children were more likely to agree (i.e., to conserve the length) in this task than in the traditional task.

Other researchers reported that children do not necessarily need to develop conservation before they can learn measurement ideas (Clements, 1999b; Hiebert, 1981; Pettito, 1990). From kindergarten to Grade 2, most children significantly improve in measurement knowledge (Ellis, Siegler, & Van Voorhis, 2003). They learn to represent length with a third object by using transitivity to compare the length of two objects that they cannot directly compare in a variety of contexts (Hiebert, 1981). They can also use given units to find the length of objects and associate higher counts with longer objects (Hiebert, 1981, 1984). Nunes and Bryant (1996) found that some 5-year-olds, but most 7-year-olds, can make inferences about the relative size of objects by using their concept of a unit; for example, when measuring two objects, if the number of units is the same, but the units are different sizes, then the sizes of the objects must be different. Even the 7-year-olds, however, found such tasks demanding and found conversion among units challenging.

In their work with children in Grades 3 and 5, Saxe et al. (2007) suggested that children at both ages (9 and 11 years old) encountered difficulties when researchers asked them to name the location of given fractions along a nonroutine number line divided into intervals of unequal lengths (e.g., a number line showing the interval from 0 to 1, with the eighths marked in the first half and quarters in the second half). However, children's difficulties decreased with age. In another study, Izsák, Tillema, and Tunç-Pekkan (2008) asked children in a sixth-grade classroom to use fractional number lines to model addition and subtraction operations with fractions. These children demonstrated subtle differences in the ways that they represented and coordinated the iteration of a subinterval from left to right through an interval. Likewise, when children were partitioning an interval, researchers identified subtle differences in the children's responses. Children's ways of iterating and partitioning both appeared to influence their understanding of the teacher's explanations.

Researchers have performed much work on the design of the ruler as a length-measurement tool. Students can, and often do, use rulers correctly (i.e., align the object to the end of the ruler and read the number from the end) without understanding. Lehrer and Shauble (2000) posited that the design for effective and efficient use in unconstrained situations often conceals children's misconceptions. Mariotti (2002) claimed that we hand children what we see as a tool, the ruler, but the ruler is simply an artifact in the hands of children. Confirmation of this fact often occurs when educators ask children to find the length of an object by using a ruler that has no numeral labels or when children cannot align the object that they are measuring with the zero point of the tool (Bragg & Outhred, 2004). The misunderstanding of this artifact is often a result of the confusion about how the tool displays the units. When educators ask children what they should count on the ruler when measuring, many children indicated the hash marks rather than the intervals between those marks (Bragg & Outhred, 2004; cf. Cannon, 1992).

Many suggestions from research describing how educators can best help children understand the ruler as a meaningful tool focus on explicit attention to unit identification. Bragg and Outhred (2004) suggested using motion to draw attention to the interval. They recommended that children count while they sweep a finger through the space from one hash mark to the next in an attempt to help children see the unit markers as "the point where each unit starts and ends" (Bragg & Outhred, 2004, p. 164). Support for this suggestion came from Cullen and Barrett (2010), who found that children using a *sweeping-the-interval* strategy produced correct responses 100% of the time in tasks in which the children could not align the object with the zero point of the ruler. Cullen and Barrett reported two additional effective strategies, *point to midpoint* and *spanning with fingers*, and concluded that the commonality "among the three effective strategies is that they encourage modelling of the individual unit segments along the object" (p. 287).

In summary, children have an intuitive understanding on which they can base reasoning about distance and length, but that reasoning develops over years, possibly requiring specific educational experiences to build concepts and skills that enable children to gain competence with length measurement. These concepts and skills include the ability to (a) align endpoints (Piaget et al., 1960), (b) use a third object and transitivity to compare the length of two objects that cannot be compared directly (Hiebert, 1981), (c) place length units along objects to find their lengths and associate higher counts with longer objects (Hiebert, 1981; Stephan & Clements, 2003), (d) understand the need for equal-length units and the notion of subdividing or partitioning an object into equal-length units (Ellis et al., 2003) and then further subdivide units into fractional units, (e) make inferences about the relative size of objects (e.g., if the number of units are the same and the units are different, then the total size is different; Nunes & Bryant, 1996), and (f) use rulers accurately and meaningfully (Lehrer, Jaslow, & Curtis, 2003; Stephan & Clements, 2003).

INITIAL LEARNING TRAJECTORY FOR LENGTH MEASUREMENT

A review of the preceding studies, as well as others, formed the basis for the initial LT for length measurement, shown in Figure 2.1. The first column describes the behaviors indicative of each level, with examples of observable behaviors (indented). The second column describes the mental actions on objects that we hypothesized underlie those behaviors (for full details, see Clements & Sarama, 2009; Sarama & Clements, 2009b). The developmental progression of the LT for length measurement presented in Figure 2.1 was the starting point for the research reported in this monograph.

We describe instructional tasks, a component of the LT for length measurement, throughout the manuscript. The focus of the longitudinal study reported in this monograph was not on elaborating and testing specific tasks; thus, the figures presenting the initial LTs (i.e., Figures 2.1, 3.1, and 7.1) do not include them. We are currently focusing on instructional tasks and teaching in a new National Science Foundation–funded project.

In the initial LTs as shown in Clements and Sarama (2009) and Sarama and Clements (2009b), the first column was age, because our initial research review synthesized many studies that were not developmental, and the age of the children in those studies helped us connect their results. However, as explained there, age is simply a noncritical correlate of the development of the important levels of thinking.

One purpose of this study was to test, refine, and extend the initial LT for length measurement (as well as LTs for area and volume measurement). For both the initial and revised LTs, we emphasize the developmental progressions, including the hypothesized mental actions on objects and schemes that account for the behaviors indicative of the levels of thinking. *Schemes* are cognitive units with three components: a structure that recognizes a situation, a process that acts on that situation, and a result.

Developmental Progression	Actions on Objects
Pre-Length Quantity Recognizer. Does not identify length as attribute. "This is long. Everything straight is long. If it's not straight, it can't be long."	Action schemes[a], both physical/kinesthetic and visual, implicitly trace linear extent—initially does not connect with explicit vocabulary and later only partially connects with vocabulary.
Length Quantity Recognizer. Uses appropriate vocabulary to identify length or distance as attribute. May understand length as an absolute descriptor (e.g., all adults are tall), but not as a comparative (e.g., one person is taller than another). "I'm tall, see?"	Action schemes connect with length vocabulary, and child uses them to compare lengths. Salient differences at one end of the objects can substitute for a scan; however, this scheme may lead to inaccuracies if the other endpoints do not align.
Length Direct Comparer. Physically aligns two objects to determine which is longer or whether they are the same length. Stands two sticks up next to each other on a table and says, "This one's bigger."	The scheme addresses length as a linear extent from endpoint to endpoint of a path. With perceptual support, children can mentally, then physically, slide and rotate objects into alignment and compare their endpoints.
Indirect Length Comparer. Compares the lengths of two objects by representing them with a third object. Compares length of two objects with a piece of string. Note: At this level, children are also aware of the process of measurement; however, when asked to measure, may assign a length by guessing or moving along a length while counting (without equal-length units) or using a ruler (but lacking in concepts and skill). Moves finger along a line segment, saying 10, 20, 30, 31, 32.	Can build, maintain, and manipulate a mental image of a particular length. With perceptual support, can compare such images. For some, may apply explicit transitive reasoning to the images or their symbolic representations. If someone asks the child to measure, the counting scheme operates on an intuitive unit of spatial extent or amount of movement, directing movements along a length while counting. The sensory-concrete mental actions require the perceptual support of the object that the child is measuring.

Figure 2.1 continues on next page

Developmental Progression	Actions on Objects
Serial Orderer to 6+. Orders lengths, marked in 1 to 6 units. (This level develops in parallel with End-to-End Length Measurer) Given towers of cubes, puts in order, 1 to 6.	Scheme organized in a hierarchy, with the higher order concept an image of an ordered series (possibly implicit). A scheme that considers each object in such a series to be longer than the one before it and shorter than the one after it (resulting in a more efficient strategy) eventually complements the ability to estimate relative lengths (driving a trial-and-error approach).
End-to-End Length Measurer. Lays units end to end. May not recognize the need for equal-length units or the need to avoid gaps. The ability to apply resulting measures to comparison situations develops late in this level. Can "read off" measures from a ruler, but only reads accurately with substantial guidance. Lays 9-inch cubes in a line beside a book to measure how long it is.	An implicit concept that lengths can consist of repetitions of shorter lengths underlies the scheme (which must overcome previous schemes that use continuous mental processes to evaluate continuous extents, and thus are more easily instantiated.)
Length Unit Relater and Repeater. Measures by repeated use of a unit (but initially may not be precise). Relates size and number of units. "If you measure with centimeters, not inches, you'll need more, because each one is smaller." Iterates a single unit to measure. Recognizes that different units result in different measures and that identical units should be used. Uses rulers with minimal guidance.	Action schemes include the ability to iterate a mental unit along a perceptually available object. Can maintain the image of each placement while moving the physical unit to the next iterative position. With the support of a perceptual context, scheme can predict that fewer larger units will be necessary to measure the length of an object. These action schemes allow the application of counting-all addition schemes to apply to measures.

Figure 2.1 continues on next page

Developmental Progression	Actions on Objects
Length Measurer. Measures, knowing need for identical units, relationship between different units, partitions of unit, zero point on rulers, and accumulation of distance. Begins to estimate. "I used a meter stick three times, then there was a little left over. So, I lined it up from 0 and found 14 centimeters. So, it is 3 meters, 14 centimeters in all."	The length scheme has additional hierarchical components, including the ability to simultaneously image and conceive of an object's length as a total extent and a composition of units. This scheme adds constraints on the use of equal-length units and, with rulers, on use of a zero point. Units themselves can be partitioning, allowing accurate use of units and subordinate units.
Conceptual Ruler Measurer. Possesses an "internal" measurement tool. Mentally moves along an object, segmenting it, and counting the segments. Operates arithmetically on measures ("connected lengths"). Estimates with accuracy. "I imagine one meter stick after another along the edge of the room. That's how I estimated that the room's length is 9 meters."	Interiorization of the length scheme allows mental partitioning of a length into a given number of equal-length parts or the mental estimation of length by projecting an image onto objects (present or imagined).

[a] A scheme is a cognitive unit with three components: a structure that recognizes a situation, a process that acts on that situation, and a result.

Figure 2.1. Initial developmental progression component of the LT for length measurement (from Sarama & Clements, 2009b, covering only birth through the primary grades).

EVALUATING, REFINING, AND EXTENDING THE INITIAL DEVELOPMENTAL PROGRESSION: QUALITATIVE RESULTS OF TEACHING EXPERIMENTS

We have reported elsewhere on the validity and ways of implementing and improving the initial developmental progression of the LT for length measurement by describing the progress of this cohort of children with linear measurement tasks during the first 2 years of this study (Barrett et al., 2011; Barrett et al., 2012; Sarama, Clements, Barrett, Van Dine, & McDonel, 2011). This section briefly summarizes some of those findings.

Pre-Kindergarten to Grade 2 Northeast Cohort

Our investigation of the use of the initial LT for length measurement showed that although children from pre-K to Grade 2 have the ability to understand and learn length concepts and skills, this development is a slow and extended process. Hence, in general, the children in the Northeast cohort exhibited behaviors and mental actions on objects consistent with the initial developmental progression of the LT for length measurement.

Length Quantity Recognizer. During pre-K, the children demonstrated thinking and behaviors indicative of the Length Quantity Recognizer level during the initial assessment. For example, Alice consistently used the word "long" to describe objects that extended further than others along one dimension; however, she was not yet able to compare lengths. Similarly, Marina put a string up next to a doll and said that the doll (which was shorter than the string) was "longer."

Length Direct Comparer. All the children in the Northeast cohort moved from the Length Quantity Recognizer level to the Length Direct Comparer level. For example, in the spring of her pre-K year, we asked Edith which of four wooden craft sticks was the same length as a strip of paper. She placed the strip next to each stick, in turn, lining up the endpoints, and said, "Not this one . . . not this one . . . not this one . . . This one!" Similarly, when we asked Alice to describe and compare several animals in a picture, she correctly used vocabulary and hand gestures to describe and indicate length (for more details on Alice's thinking and behaviors on this task, see Sarama, Clements, Barrett, et al., 2011, p. 672).

End-to-End Length Measurer. Children next progressed to the End-to-End Length Measurer level, placing multiple units along the object to be measured. At this level, children initially left gaps—although they improved from the previous level in placing the units contiguously—and then counted these individual units to report the length. Some children achieved this level during pre-K and others achieved it during kindergarten. As an example, in the spring of pre-K, Edith laid blocks end to end with no gaps around a rectangle and counted them to measure the perimeter. Similarly, in the spring of kindergarten, Marina laid rulers end to end along the edge of a desk and counted the number of rulers.

Length Unit Relater and Repeater. Our finding that even confident End-to-End Length Measurers could not or would not measure if they lacked a sufficient number of units to cover the length of the object substantiated the separation of the End-to-End Length Measurer level from the Length Unit Relater and Repeater level. Indeed, this level appeared to be a difficult one to achieve. Only two children showed consistent signs of the ability to iterate a single unit to measure a length or to use a ruler with full understanding, both behaviors indicating the Length Unit Relater and Repeater level, by the end of the third year of the study (Grade 1). That is, the researchers gave children a ruler and inch tiles and asked them, "How many inches long is the ruler?" All the children in the Northeast cohort could line up the tiles and count them to show that the ruler was 12 inches long. However, when the researcher pointed to the 5 on the ruler and asked the number of inch tiles from the 0 to the 5 on the ruler, Edith and Alice were the only children at the Northeast site who could relate the number on the ruler to the number of tiles that would span the length from the zero point on the ruler to a number on the ruler, as indicated by their report of the answer 5 without counting tiles.

Summary of findings for the Northeast cohort. Researchers have long recognized the difference between the nonmeasurement (nonnumerical) processes of direct and indirect comparison and the numerical measurement processes (e.g., Hiebert, 1981; Piaget et al., 1960). However, some developmental progressions consider them separate or distinct processes, whereas others, such as ours, consider them to be different but related strategies (Barrett & Battista, 2014). We confirmed that all the children at the Northeast site used direct comparison before they used other nonnumerical and numerical processes. Our findings therefore suggest that the actions on objects indicative of the Length Direct Comparer level, such as alignment and comparison of the endpoints of lengths, related hierarchically to all subsequent length development; that is, more sophisticated levels depended on the actions on objects developed at the Length Direct Comparer level. That finding supported our view of related strategies.

However, the Indirect Length Comparer level of thinking did not fall in line with the initial developmental progression of the LT for length measurement, so we made revisions. First, we renamed it *Length Indirect Comparer* for more grammatical consistency with the other levels. Second, although children demonstrated progress on indirect length-comparison competencies, that progress was neither complete nor reliable. None of the children used indirect comparison in pre-K, even those who effortlessly used direct comparisons. Even later, when they showed signs of using it, they were not consistent. Unlike in previous studies, we assessed these competencies in multiple contexts and found that children vacillated frequently between using a third object with transitive reasoning and simply making visual comparisons. Alice, for example, used a third object to compare when the first and second objects were horizontal and vertical but not when they were oblique (Sarama, Clements, Barrett, et al., 2011, p. 673). Further, we did not find that children used indirect comparison at a specific point in their progression through the LT. For example, all but two of the children reliably laid units end to end to measure an object before they reliably used indirect comparison with transitive inference.

Children's use of visual estimation (and, we assume, mental imagery of the objects) to conduct length comparisons initially suggested that such behaviors might signify thinking at the Length Indirect Comparer level: We thought it was possible that they were using a stable mental image as the third object. However, observations indicated that such comparisons were more consistent with a mental application of the pattern of thinking at the Length Direct Comparer level. That is, the mental image appeared to be a straightforward representation of the first object, which the children directly compared with the second object with no transitive inference involved. We hypothesized that such mental imagery serves as a transition to the development of indirect comparison with explicit transitive reasoning. We also posited that such results are consistent with our understanding of reasoning about early perceptual comparisons of discrete quantities (for more on this argument, see Sarama, Clements, Barrett, et al., 2011, pp. 678–679). This theorizing was post hoc, and researchers must conduct further evaluation in future research.

Children also used visually based strategies in measurement tasks by using either repeated pointing or eye movements (saccades) to move along an object while counting. We began by hypothesizing this strategy as indicative of the Length Unit Relater and Repeater level, in which children iterated an internalized stable unit of length. However, focused subsequent observations provided disconfirming evidence, showing that such inaccurate iteration appeared to be an unsophisticated instantiation of behaviors at the End-to-End Length Measurer level. Children did not make or even appear to make an effort to maintain equal-interval units. Therefore, the role of visual strategies and mental imagery in children's development of length measurement is a second issue for future research, the goal of which is to confirm or disconfirm and to elaborate on our nascent findings and conclusions. (Such developments do play the hypothesized role in more sophisticated levels, such as the Conceptual Ruler Measurer.) In conclusion, children reliably move through the levels posited in the initial LT, except for the Length Indirect Comparer level, which we hypothesize develops in parallel or integrated with measurement levels (End-to-End Length Measurer and more sophisticated levels).

Grade 2 to Grade 5 Midwest Cohort

In our work at the Midwest site, we investigated the use of the initial LT for length measurement (see Figure 2.1) to describe children's thinking about linear measurement from Grade 2 through Grade 5. The main conclusion was that evidence supported our previous claims about the hierarchical nature of the existing levels of the initial LT for length measurement (Sarama & Clements, 2009b). However, we found that the children in the Midwest cohort also exhibited thinking about length measurement that was beyond the scope of the levels described in the initial LT for length measurement. The following sections describe each level in succession. All the children in our Midwest cohort progressed consistently through the levels, as shown in the growth charts provided subsequently in this chapter.

End-to-End Length Measurer. Children in the Midwest cohort also demonstrated thinking and behaviors indicative of the End-to-End Length Measurer level. One of the eight children, Sara, began the study as an End-to-End Length Measurer, but during Grade 2 she progressed quickly to the Length Unit Relater and Repeater level (Barrett et al., 2012). On the initial assessment, Sara used a short paper strip as a unit and measured the length of a longer object by moving the strip along the longer object while counting. However, instead of always moving the short paper strip the length of a paper strip, each successive move was shorter than the length of the paper strip, thereby creating overlaps. In a follow-up to the initial assessment during the spring of Grade 2, the researchers asked Sara to use a 1-in. yellow strip to measure a 5-in. strip. Sara reported three different measures (i.e., 10, 11, and 14 inches) on her three attempts to measure this strip. The different measures resulted from her inconsistency in maintaining unit size in attempting to iterate the yellow strip. However, Sara correctly reported the length of various strips ranging in length from 4 to 7 inches when given enough yellow strips to cover the strip that she wanted to measure (see Barrett et al., 2012, p. 38, for more information on Sara and subsequent tasks). In contrast to Sara, the other children at the Midwest site demonstrated thinking more sophisticated than the End-to-End Length Measurer level from the onset of the study during Grade 2.

Length Unit Relater and Repeater. The transition from the End-to-End Length Measurer level to the Length Unit Relater and Repeater level is crucial because children begin to use less direct observation and activity and instead use more imagery and mental actions on objects. Children decrease their reliance on perception and immediate activity with physical objects to span or cover an entire linear object with a complete collection. During this transition, they begin to recognize that they can move linear objects, that they can scan along a part of a linear object (i.e., sweeping through a segment), and that a set of repeated placements of an object constitutes the length of another longer object.

At the Northeast site, only two children showed thinking indicative of the Length Unit Relater and Repeater level. However, at the Midwest site, six of the eight children began the study as Length Unit Relaters and Repeaters. Children at the Length Unit Relater and Repeater level are able to count intervals by counting tick marks (productively or unproductively), or they may read off number labels along a ruler to determine the length of a line segment. The integration of schemes for sweeping, marking endpoints, and coordinating the counting activity of ordinal and cardinal sequences all constitute critical actions on mental objects at the Length Unit Relater and Repeater level.

For example, Anselm demonstrated thinking at the Length Unit Relater and Repeater level throughout Grade 2 and into Grade 3. Anselm struggled longer than other children with the problem of confusing points or hash marks for indicators of linear units, but we also found that he was able to see and explain the alternative (e.g., that we find units of length between hash marks on a ruler). To challenge his misconception, we asked Anselm to draw a 3-in. line segment. Anselm drew a 2-in. line segment but asserted that it was 3 inches long. Then we asked him to draw a 2-in. line segment, and he drew a 1-in. line segment (and asserted that it was 2 inches long). When we asked him to draw a 1-in. line segment, Anselm also drew a 1-in. line segment. He found it "really weird" that his 1-in. line segment and his 2-in. line segment were the same length. After additional tasks, Anselm seemed to overcome this misconception (see Barrett et al., 2012, pp. 42–43, for more details on Anselm's misconception and the additional tasks) and asserted that he should "count the spaces" rather than the hash marks.

After overcoming this misconception, Anselm appeared to exhibit unusually swift growth, which we saw as allowing him to catch up to the cohort of children in this study. We take this growth as evidence that his schemes for iterating units, partitioning linear objects, and sweeping through intervals along linear objects remained partially integrated because of his attempts to reconcile accounts of prior strategies for counting points and for counting intervals within the context of the teaching-experiment episodes. After he determined that point counting must not take priority over the sweeping of intervals that constituted units of length, he progressed quickly through the simple iteration of units (i.e., the Length Measurer level) into more abstract levels.

Length Measurer. At the Length Measurer level, children are able to simultaneously image and conceive of an object's length as a total extent and a comparison of units. Children see length as a ratio comparison between the unit and the object measured. They consistently measure straight paths, use equal-length units, understand the zero point on the ruler, and can partition units to use units and subunits for the purpose of increasing precision. Children operating at this level sometimes make rounding errors when measuring each segment along a bent path formed of several segments, and thus they may not equate the sum of the parts of the bent path to the length of the whole.

The children in the Midwest cohort showed evidence of thinking at the Length Measurer level as early as the initial assessment administered in the spring of Grade 2 but only on some of the assessment items: They more often exhibited thinking at the Length Unit Relater and Repeater level on that assessment. By the end of Grade 3, all the children were predominantly using thinking at the Length Measurer level, with occasional fallbacks to the Length Unit Relater and Repeater level, which is consistent with our theoretical perspective hierarchic interactionalism.

For example, Danny exhibited thinking at the Length Measurer level during the spring of Grade 3 when the researchers presented a fixed-perimeter task. We provided him with a 12-in. looped string with knots to demarcate the units and asked him to explore the set of rectangles with a perimeter of 12. He used the string to make a long skinny rectangle and said he thought that it was 1 cm on each short side and about 5½ in. on each long side. When the researchers asked him to make a record (i.e., a drawing) of his rectangle, he drew a rectangular region with tick marks along the bottom to represent the knots on the string, as shown in Figure 2.2.

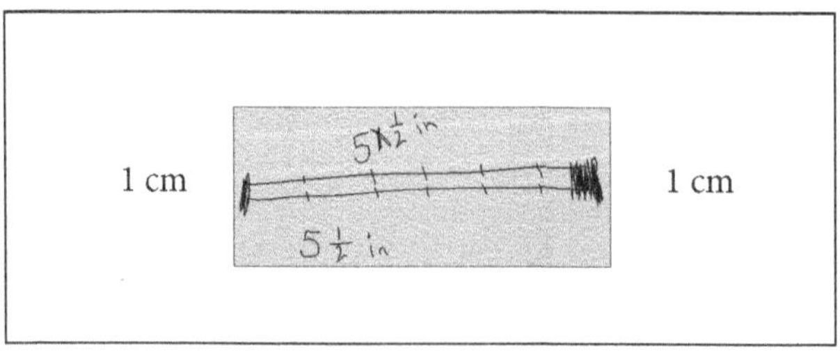

Figure 2.2. Danny's record of a 5½ in. × 1 cm rectangle.

Danny labeled the long side as 5½ in. and the short side as 1 cm. He explained that the two long sides of his rectangle added to 11 in. and that he had 1 in. left for the short sides, so each short side should be 1 cm. We conjecture that although Danny was comfortable referring to a side longer than 5 in. as 5½ in., he wanted to avoid fracturing a single unit (cf. Hirstein, Lamb, & Osborne, 1978; Kamii & Kysh, 2006). Further, we note that 1 cm is approximately ½ in. Danny's strategy for determining side lengths of 5½ in. and 1 cm from a perimeter of 12 in. is consistent with the Length Measurer level because the length of his bent path (the rectangle with a perimeter of 12 in.) is not quite the sum of its parts (the side lengths were 5½ in. and 1 cm as he recorded them) because of the error introduced by changing the unit of the short side from inches to centimeters.

Conceptual Ruler Measurer. At the Conceptual Ruler Measurer level, children possess an "internal" measurement tool (Steffe, 1991). That is, they employ explicit strategies to estimate lengths reasonably, such as mentally iterating internal units of length or partitioning a length into equal-length parts. Children at the Conceptual Ruler Measurer level project or translate given lengths to determine missing lengths, and they notice or are perturbed by geometric inconsistencies. At this level, children increasingly use multiplicative reasoning in comparison situations.

All the children in the Midwest cohort began to show some evidence of thinking at the Conceptual Ruler Measurer level by the spring of Grade 3 and evinced thinking at the Conceptual Ruler Measurer level by the end of Grade 4. Four children—Danny, Arielle, Drew, and Owen—consistently demonstrated Conceptual Ruler Measurer level thinking throughout Grade 4. However, two children—Abby and Anselm—relied on thinking at the Conceptual Ruler Measurer level for a short time early in Grade 4 before showing growth toward the next level of the LT for length measurement.

Danny and four other children first exhibited thinking at the Conceptual Ruler Measurer level during the spring of Grade 3 when researchers asked them to estimate the length of the wall (in feet) of a room where the researchers conducted teaching episodes. The room measured 17½ feet long. Danny scanned the length of the room by slowly turning his head and incrementally pausing before scanning further. This behavior seemed to indicate his mental iteration of an internal unit. After scanning the length of the room in this way, Danny accurately answered "seventeen and a half." When the researcher asked him how he thought about his answer he said, "I just know that the size of a foot . . . it's that big (indicating his estimate for a foot with his hands), and I just placed it every time . . . as much as the real size." Another child, Drew, gave a reasonably accurate estimate of 17 ft. He explained that he knew that he was 4 ft. 7 in. tall and he imagined himself standing against the wall. Drew's use of a conceptual unit of units (a rotated projection of his body length that he could relate to feet and inches) and Danny's application of a conceptual standard unit (the iterated mental image of a foot) are both consistent with thinking at the Conceptual Ruler Measurer level.

Three other children—Anselm, Abby, and Owen—understated the measurement of a room length but were able to impose mental units onto the room length with reasonable success. Anselm initially answered 15 feet but changed his answer to 10 feet. He explained that he thought about the swimming pool in his school that was 10 feet deep and used this reference to make a benchmark comparison and determined that the room was 10 feet long. Abby answered 11 feet; she explained that she imagined moving a foot-long sandwich along the length of the room. Owen answered 4 yards and explained that he pictured a yardstick. Anselm's strategy of comparing the length of the room to a rotated projection of a benchmark and Abby and Owen's strategy of mentally iterating a conceptual standard unit that they associated with specific objects are also consistent with thinking at the Conceptual Ruler Measurer level even though their estimates were about 30% to 40% too short.

During the spring of Grade 3, children participating in the study began to demonstrate the use of behaviors at the Conceptual Ruler Measurer level when resolving fixed-perimeter tasks. For example, when the researchers gave Danny a 12-in. looped string to use as a tool and asked him to make a rectangle with it, he used it to make a 5 × 1-in. rectangular shape. When the researchers asked him to make a record of the rectangle that he had made with the string, he sketched a rectangle and labeled the long side as 5 in. and the short side as 1 in. (see Figure 2.3).

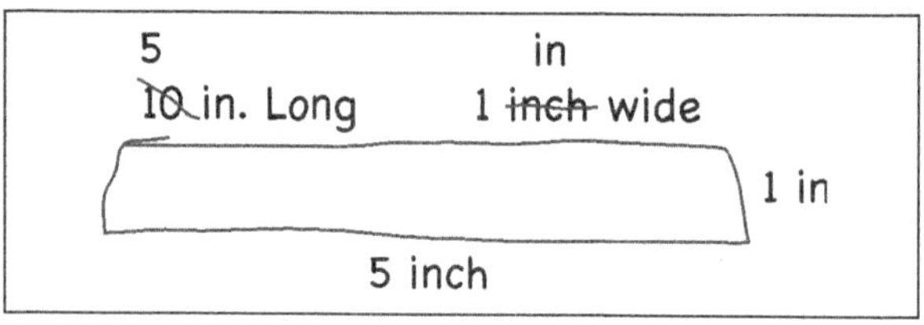

Figure 2.3. Danny's record of a 5-in. × 1-in. rectangle.

Danny's ability to connect his figural image (i.e., drawing the 5-in. × 1-in. rectangle) and perceptual image (i.e., making a rectangle from the 12-in. looped string), as well as his recognition of geometric consistencies (e.g., opposite sides of a rectangle are congruent) within these images, is evidence of thinking at the Conceptual Ruler Measurer level.

During the fall of Grade 4, Abby successfully resolved a missing-measures task by projecting or translating given lengths to determine missing lengths. The researcher showed Abby the image of a bent path, depicted in Figure 2.4, and asked, "If we walk along a path from *A* through *B* to *C*, how long would that path be?"

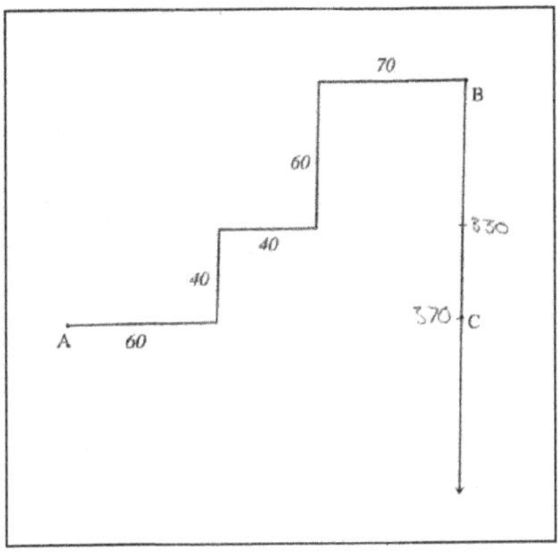

Figure 2.4. Abby's application of a missing-measures strategy.

Without pointing to the path, Abby correctly counted, saying, "100, 200, 270, . . . , 330, . . ." The researcher then interrupted Abby by handing her a pen and asking her to mark the 330. She touched the pen to one end of the vertical 60-unit segment and then swept her pen over to $\overline{BC}$, which indicates that she projected this vertical 60-unit segment over to $\overline{BC}$. She then repeated this action by touching one end of the vertical 40-unit segment saying, "that would be 40 so . . . 370" while she swept her pen over to point *C* on $\overline{BC}$, correctly labeling it as 370. Abby's strategy of projecting or translating the vertical 60- and 40-unit segments over to $\overline{BC}$ to determine the total length of the path from point *A* through *B* to *C* is consistent with thinking at the Conceptual Ruler Measurer level.

More sophisticated levels than Conceptual Ruler Measurer. The transition into thinking more sophisticated than the Conceptual Ruler Measurer level marks a shift in the focus of the trajectory from the conceptual and procedural aspects of length measurement toward the confluence of three components of measurement thinking: linear quantity, geometric or spatial characteristics, and reasoning toward proof (cf. Barrett, Clements, Klanderman, Pennisi, & Polaki, 2006). On the basis of the data from the Midwest cohort, we identified two levels more sophisticated than the Conceptual Ruler Measurer level that we had not included in the initial developmental progression of the LT for length measurement. We later called these two levels ("Level *n* + *1*" and "Level *n* + *2*") the *Integrated Conceptual Path Measurer level* and the *Abstract Length Measurer level*, respectively. We used data from the Midwest cohort to clarify the definition of these two levels. For the remainder of this chapter, we use the final names assigned to these levels to avoid confusion.

Integrated Conceptual Path Measurer. At the Integrated Conceptual Path Measurer level, children integrate and compare sets of units along each section of a bent path. When they reflect on the measure of a bent path or the perimeter of a polygon, they regard a group of units as a flexible object, a "string" of units wrapped around the entire perimeter or along the entire path. Therefore, in the context of a fixed-perimeter or fixed-path-length task, children at the Integrated Conceptual Path Measurer level are able to compensate for changes made to one side of a figure by adjusting other sides to maintain the fixed overall length. Although they can find several related cases of polygons with the same perimeter, they may not yet be able to organize and synthesize a set of related polygons on the basis of perimeter to formulate and justify a valid argument. At this level, children also begin to coordinate other measures with linear measures, such as angle or curvature, and show well-developed ideas about precision, such as constructing smaller units for the purpose of increasing precision.

All the children in the Midwest cohort showed evidence of thinking at the Integrated Conceptual Path Measurer level by Grade 4. During that time, two children—Drew and Owen—sporadically used thinking at the Integrated Conceptual Path Measurer level. Danny and Arielle relied on thinking at the Integrated Conceptual Path Measurer level intermittently in some task situations. In contrast, Abby and Anselm both consistently showed thinking at the Integrated Conceptual Path Measurer level.

Danny and Abby demonstrated thinking at the Integrated Conceptual Path Measurer level in Grade 4 within the context of approximating the length of a curve. When researchers asked Danny to compare the length of a pen to the length of a curve, he placed the pen tangent to the curve and rotated it while moving it along the path, stopping at different points of tangency (see Figure 2.5).

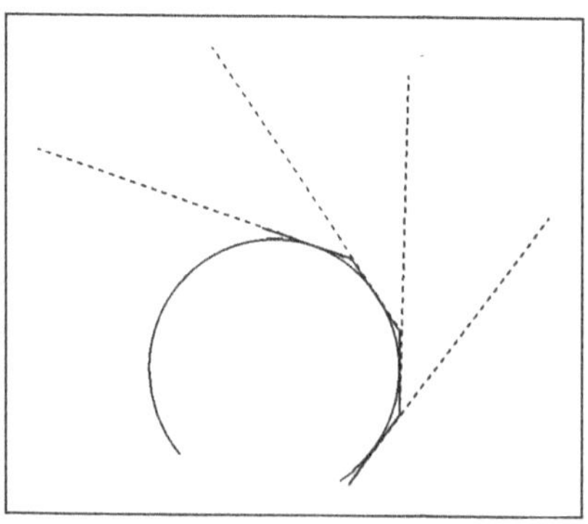

Figure 2.5. A computer-generated representation to illustrate Danny's subdividing approach on the curved-path task. The dashed line segments represent pen-length iterations.

He then said, "I'm thinking it might be two or three" and explained that he was thinking about subdividing the pen length (represented by the dashed line segments in Figure 2.5) into fourths and iterating a fourth (represented by the solid line segments in Figure 2.5) tangentially along the curve. Danny said, "This is about one fourth of it, two fourths, three fourths, and four fourths" [moving the pen to a new point of tangency with each count]. He then noted another set of four fourths along the other half of the curve and said that the curve was two pens long. Abby used a similar approach when approximating the length of the same curve by attending to tangents while using an incremental adjustment method. She incrementally moved the pen along the curve by changing its position of tangency with one hand while accumulating length along the pen with her other hand until she used a full pen length. At this point, she was halfway around the curve and said it was "two times bigger than this [the pen]." Danny's subdividing approach and Abby's incremental adjustment method indicate that each constructed smaller units to coordinate curvature with linear extent.

Anselm first showed evidence of thinking at the Integrated Conceptual Path Measurer level during the fall of Grade 4. The researcher presented him with a 24-unit looped string and asked him to make a rectangle with it. Anselm used the string to make a 2 × 10 rectangle. When the researcher asked him to make a record of it, he drew a rectangle and told the researcher that it was 24 units "around here" while he traced his finger around the boundary of the rectangle to indicate the perimeter. Anselm correctly labeled the short sides as 2 and the long sides as 10. The researcher then asked Anselm to make another rectangle with a perimeter of 24, and he manipulated the string to make a new rectangle that he said was 3 on the short sides. Without returning to the string, he labeled the short sides as 3 and the long sides as 9. Anselm explained that he knew that the 3 × 9 rectangle had a perimeter of 24. He said that he subtracted one unit from each long side of the 2 × 10 rectangle and added that unit to each short side to form the 3 × 9 rectangle. This strategy is consistent with thinking at the Integrated Conceptual Path Measurer level because Anselm was able to compensate for changes made to one side of the 2 × 10 rectangle (i.e., increasing the length of each of the short sides by one unit) by adjusting other sides (i.e., decreasing the length of each of the long sides by one unit) to maintain the fixed perimeter.

Abstract Length Measurer. The most sophisticated level of the developmental progression of the revised LT for length measurement is the Abstract Length Measurer level. The Integrated Conceptual Path Measurer and Abstract Length Measurer levels both emphasize the development of increasingly sophisticated logical thought, reflection, explanation, and justification in geometric situations. However, the Abstract Length Measurer level constitutes an advance over the Integrated Conceptual Path Measurer level because it generalizes the schemes to apply to multiple cases of complex paths (e.g., perimeters of polygons), including comparing and sequencing their lengths (i.e., perimeters).

At this level, children engage dynamic imagery to coordinate and operate internally on collections of units of units as well as collections of complex paths. Although they can find several related cases of polygons with the same perimeter, they may not yet be able to organize and synthesize a set of related polygons based on perimeter to formulate and justify a valid argument. Children at this level coordinate additive and multiplicative reasoning in fluent ways. They also engage in proportional reasoning about coordinated cases of paths and recognize and reflect on patterns among cases.

Three children—Drew, Anselm and Arielle—used thinking at the Abstract Length Measurer level on a sequence of fixed-perimeter tasks in Grade 4. During the fall of Grade 4, the researcher gave Drew a 24-unit looped string and asked him to explore the set of rectangles with a perimeter of 24. He discovered 8 × 4, 11 × 1, and 9 × 3 rectangles with the string. In each case, he counted the number of units on perpendicular adjacent sides, added side lengths to verify that the perimeter was 24, and made a record of each rectangle. Drew then made a fourth rectangle without the string. He drew a rectangle and explained that he was going to make the short side longer than the short side of the 8 × 4 rectangle and labeled opposite short sides as 5 and the long sides as 7. When designing a fifth rectangle, Drew noticed a recursive rule for finding the side lengths of a new rectangle by thinking about the side lengths of the previous one. He explained that every time that he made a new rectangle by thinking about making the short side (of the previous rectangle) one unit longer, he had to subtract one unit from the long side of the previous rectangle so the perimeter would always be 24. Drew used this rule to design a new rectangle by making each of the short sides of the 5 × 7 rectangle longer by one unit and each of the long sides shorter by one unit to form a 6 × 6 rectangle. He then noted that he had rectangles with short sides of 1, 3, and 4 (as well as 5 and 6), so he decided to make a rectangle with a short side of 2 and a long side of 10. In contrast to Anselm, who created only one rectangle by making compensating changes to the length of one side of a rectangle when determining the length of another, Drew created multiple rectangles by using that rule. Hence, Drew's approach to this fixed-perimeter task is consistent with thinking at the Abstract Length Measurer level because his recursive rule for finding rectangles with a perimeter of 24 suggests that he was thinking about repeated or coordinated cases of paths and reflecting on a pattern in these cases.

Anselm showed evidence of thinking at the Abstract Length Measurer level on a related task during the fall of Grade 4. After Anselm had explored the set of rectangles with a perimeter of 24 by using a 24-unit looped string as a tool during the previous teaching episode, the researcher introduced Anselm to a corner-maker tool with a 12-unit string (see Figure 2.6) as a tool to represent "an L" of a rectangle (consisting of a short side and a long side of a rectangle).

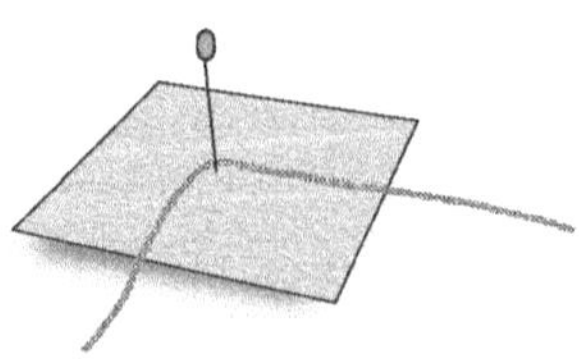

Figure 2.6. Corner-maker tool and 12-unit string.

This tool provided an opportunity for Anselm to explore his thinking with respect to building rectangles with nonintegral side lengths. Anselm used the corner-maker tool first to represent a 1 × 11 rectangle and then rotated the string around the pin to represent the 2 × 10 rectangle. When the researcher asked if there was a rectangle that he could make between the 1 × 11 and the 2 × 10, he rotated the string around the pin to form a 1½ × 10½ rectangle and then a 1¼ × 10¾ rectangle. The researcher then asked him how many rectangles he could make between the 1 × 11 and the 2 × 10. Anselm rotated the string back and forth around the pin through the second interval on the string and said, "probably like millions if you had like . . . this . . . like . . . really weird scientist guy that had like this little thing that moved it [a very tiny bit each time]." Anselm is the only child who explained that he could make millions of rectangles with a perimeter of 24, demonstrating a nascent awareness of the continuous sense of space. That is, he was developing the notion that it is possible to subdivide units and that this process of subdivision is potentially unlimited; this idea is consistent with thinking at the Abstract Length Measurer level.

EVALUATING THE INITIAL DEVELOPMENTAL PROGRESSION: QUANTITATIVE RESULTS

As described in Chapter 1, we administered two assessments to a wide range of children, ranging from pre-K to Grade 5, at the Midwest and Northeast sites. We administered an initial assessment in spring 2008, and we used it to identify children for our longitudinal study. We then administered a final assessment in spring 2011 to 258 children, including the children from our longitudinal study. We analyzed the data from this assessment to present the following results.

In the Rasch modeling, we found evidence of unidimensionality, with the measures explaining 59.9% of raw variance, an eigenvalue of 1.7, and only 3.7% of unexplained variance in the first contrast (acceptable values are above 40%, under 2.0, and under 5%, respectively). Evidence of length measurement as a unidimensional construct fulfills one of the underlying assumptions of Rasch modeling, thereby indicating that Rasch modeling is appropriate for the data. In addition, we found the mean square (MNSQ) infit for the items was between 0.76 and 1.32 and within the range of productive measurement, 0.5 to 1.5 (Linacre, 2005); thus, all items on the instrument were well aligned; this outcome supports our assumption that they are measuring the same construct—length.

As indicated in our introduction to Rasch modeling in Chapter 1, item reliability is an estimate of the replicability of item placement within the ordered list of items along the scale and relates conceptually to Cronbach's alpha. Item separation indices are estimates of the distance between items on that scale (i.e., the ability of the measure to differentiate items on the trait). Our data showed item separation and reliability at 7.92 and 0.98, suggesting that the item-difficulty hierarchy was suitable for the sample used in this study and that the assessment was able to distinguish between low-performing children and high-performing children. The item-person map indicated difficulty of the length items as slightly easy for the child sample in the assessment, yet our overall results supported the developmental progression in the initial LT for length measurement.

Our overall purpose in this section is to verify the developmental progression of the LT for length measurement by using quantitative data. We looked at each pair of items designed to assess a given level within the LT for length measurement and constructed intervals by using the item difficulty estimates of each. For example, we assessed the Serial Orderer to 6+ (SO) level using items SO-1 (predicted difficulty of −4.66) and SO-2 (predicted difficulty of −2.44); thus, the interval constructed for this level ranged from −4.66 to −2.44. Figure 2.7 depicts all items arranged with item difficulty on the *y*-axis and corresponding level intervals marked to represent the range of difficulty for each pair of items designed to assess the same level of understanding. Each interval, moving from left to right in the figure, corresponds to an increasingly more sophisticated level, as the developmental progression of the initial LT for length measurement hypothesized.

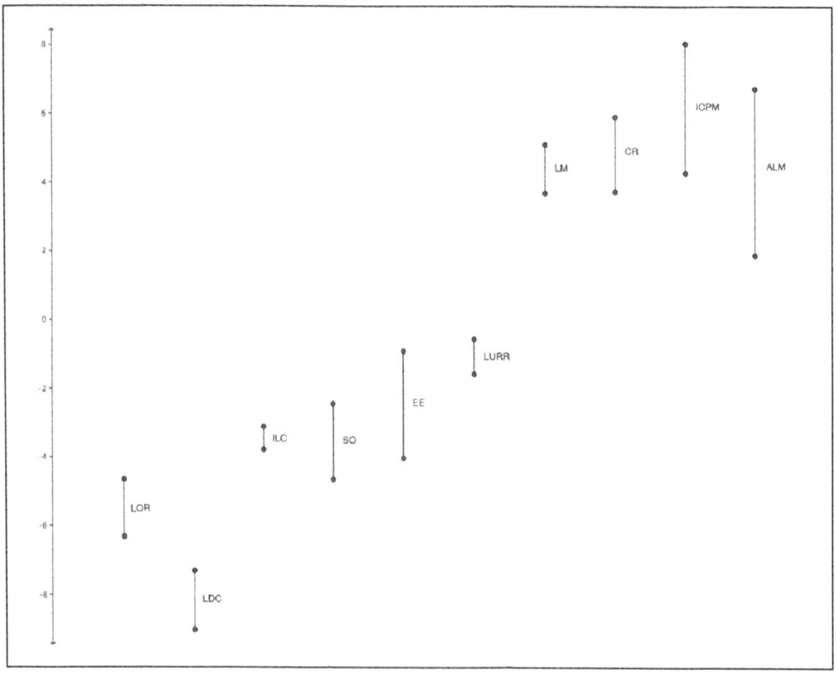

Figure 2.7. Item difficulty aligned with the levels of the LT for length measurement. LQR = Length Quantity Recognizer, LDC = Length Direct Comparer, ILC = Length Indirect Comparer, SO = Serial Orderer to 6+, EE = End-to-End Length Measurer, LURR = Length Unit Relater and Repeater, LM = Length Measurer, CR = Conceptual Ruler Measurer, ICPM = Integrated Conceptual Path Measurer, and ALM = Abstract Length Measurer.

In general, estimates indicate a progression in difficulty of the items that is generally consistent with our initial developmental progression. That is, an increase in item difficulty corresponds to a move to more sophisticated levels in the LT. The horizontal spread of these estimate displays for each item (i.e., the vertical segments) here distinguishes one hypothesized LT-level estimate from the others because otherwise the distance across is not meaningful in the display. However, the vertical distances shown by the height of each item are significant and indicate the difficulty levels spanned by each pair of items designed to assess a given level.

We identified two discrepancies in the data shown in Figure 2.7. First, the interval for the Length Direct Comparer level reaches lower than the interval for the Length Quantity Recognizer level. Content analyses of the items intended to measure thinking at the Length Quantity Recognizer level, as well as qualitative analyses of children's responses to them, indicated that they indeed targeted a higher understanding. Therefore, researchers need to develop new items to truly assess understanding at this level (most observations of the level have been informal observations of everyday activities and interactions, and they have aligned with the proposed levels). Second, the interval for Abstract Length Measurer is overall lower than that for Integrated Conceptual Path Measurer. Again, analysis of the items intended to measure children's thinking at the Abstract Length Measurer level reveals that researchers should revise or replace these items. Alternatively, this discrepancy may reveal a fault in the theoretical ordering of the levels; however, qualitative results have militated against this possibility. We discuss these qualitative results in the remainder of the chapter.

LONGITUDINAL RESULTS ACROSS LEVELS AND SITES

This section presents a collection of growth charts that we have based on data from both research sites, including one chart for each child in the study and composite charts for each cohort of children from pre-K through Grade 2 and from Grade 2 through Grade 5. These growth charts provide longitudinal accounts of the children's developing knowledge spanning 4 years. Furthermore, the composite growth charts for each cohort represent the movement through the hypothesized developmental progression of the cohort children. We interpret these charts as confirming the core of the developmental progression of the LT for length measurement, given the minimal modifications to that developmental progression that arose during the teaching experiment.

We designed the growth charts to indicate levels of thinking. The horizontal axis tracks time during each school year of the study. The vertical axis represents the hierarchical levels of the developmental progression of the LT for length measurement. A shaded rectangle indicates the dominant level of thinking identified during a teaching episode; whereas an unshaded rectangle indicates that the child was progressing to the level designated by that unshaded rectangle during a teaching episode. That is, the child was reaching into that level but was not yet dominantly at that level. Finally, a line segment indicates that the child showed evidence of thinking at that particular level on some task during the teaching episode. We do not make claims that a child is unable to reason at a particular level merely because he or she has not exhibited thinking characteristic of that level on a specific task. Instead, the use of the line segment and rectangles illustrates the range of thinking evident on a sequence of tasks during a single teaching episode. This range of thinking is consistent with hierarchical interactionalism.

Prekindergarten Through Grade 2

Edith's growth. Edith initially displayed Length Direct Comparer as her dominant level of thinking during pre-K, and this level continued during the first semester of kindergarten. At that point, her dominant level of thinking shifted to the End-to-End Length Measurer level, and that level of thinking remained dominant through the remainder of kindergarten and into the beginning of Grade 1. During the remainder of Grade 1 and Grade 2, Edith's dominant level of thinking shifted to the Length Unit Relater and Repeater level (see Figure 2.8).

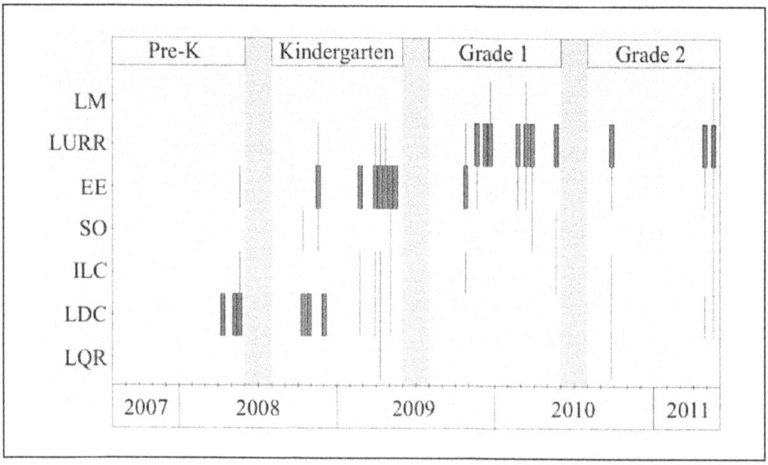

Figure 2.8. Edith's longitudinal growth chart. LQR = Length Quantity Recognizer, LDC = Length Direct Comparer, ILC = Length Indirect Comparer, SO = Serial Orderer to 6+, EE = End-to-End Length Measurer, LURR = Length Unit Relater and Repeater, and LM = Length Measurer.

Lia's growth. Lia showed evidence of thinking at the Length Direct Comparer level during both pre-K and Kindergarten 1. (Lia repeated kindergarten while in our study; we use the term *Kindergarten 1* to refer to her first year in kindergarten and *Kindergarten 2* to refer to her second.) Interestingly, evidence of thinking at the End-to-End Length Measurer level first appeared during pre-K and continued to be her dominant level of thinking throughout the 4 years of the study. For one teaching episode during Kindergarten 2, Lia's dominant level of thinking was at the Serial Orderer to 6+ (See Figure 2.9).

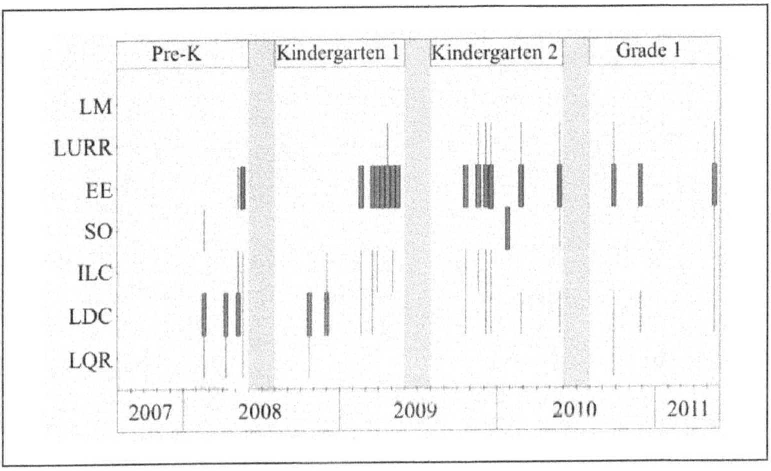

Figure 2.9. Lia's longitudinal growth chart. LQR = Length Quantity Recognizer, LDC = Length Direct Comparer, ILC = Length Indirect Comparer, SO = Serial Orderer to 6+, EE = End-to-End Length Measurer, LURR = Length Unit Relater and Repeater, and LM = Length Measurer.

Robert's growth. Robert demonstrated thinking at the Length Direct Comparer level as his dominant level during pre-K. This pattern continued during kindergarten, although the behaviors consistent with the Length Indirect Comparer and End-to-End Length Measurer levels appeared near the end of the school year. During both Grade 1 and Grade 2, we observed End-to-End Length Measurer as his dominant level of thinking. Other levels of thinking—including Length Quantity Recognizer, Serial Orderer to 6+, and Length Unit Relater and Repeater—occurred multiple times as isolated instances (see Figure 2.10).

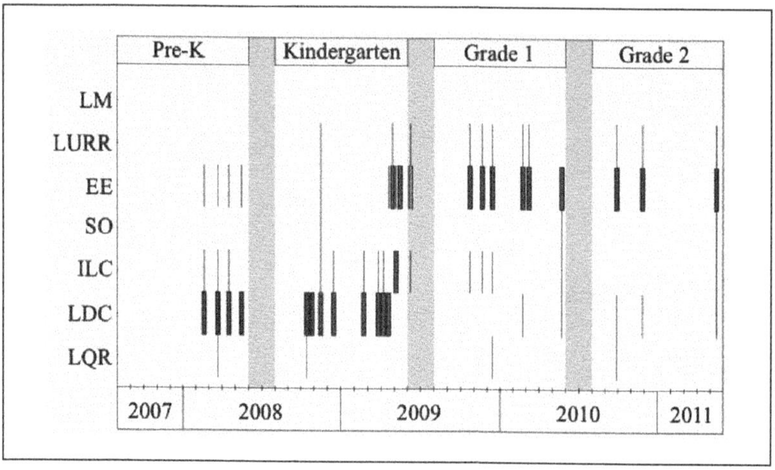

Figure 2.10. Robert's longitudinal growth chart. LQR = Length Quantity Recognizer, LDC = Length Direct Comparer, ILC = Length Indirect Comparer, SO = Serial Orderer to 6+, EE = End-to-End Length Measurer, LURR = Length Unit Relater and Repeater, and LM = Length Measurer.

Composite growth chart for prekindergarten to Grade 2. Several patterns emerge in the composite chart, shown in Figure 2.11. At the beginning of the study, children in the Northeast cohort primarily demonstrated thinking at the Length Direct Comparer level; but by the end of the study, these children had progressed to demonstrating thinking primarily at the End-to-End Length Measurer level. High variability occurred in kindergarten, including Length Direct Comparer, Length Indirect Comparer, and End-to-End Length Measurer as dominant levels. In the third year of the study (Lia's Kindergarten 2 and the other children's Grade 1), most of the dominant levels were End-to-End Length Measurer and Length Unit Relater and Repeater. This pattern continued in the fourth year for the children in this cohort (Grade 1 for Lia and Grade 2 for the other children), with occasional fallbacks to lower levels. Because children in Grade 2 often receive instruction in length measurement, anticipating that children sometimes exhibit fallbacks may be important.

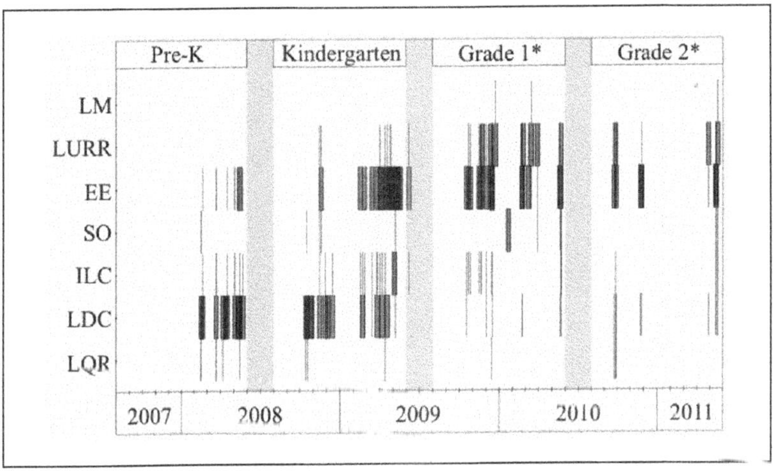

Figure 2.11. Pre-K through Grade 2 composite growth chart. Asterisks on Grade 1 and Grade 2 indicate that the Grade 1 column, the third year of the study, reports Lia's data from her second year of kindergarten, and that the Grade 2 column reports Lia's Grade 1 data. LQR = Length Quantity Recognizer, LDC = Length Direct Comparer, ILC = Length Indirect Comparer, SO = Serial Orderer to 6+, EE = End-to-End Length Measurer, LURR = Length Unit Relater and Repeater, and LM = Length Measurer.

Grade 2 Through Grade 5

Abby's growth. Abby's growth chart, shown in Figure 2.12, illustrates a steady progression, beginning with Length Unit Relater and Repeater in Grade 2 and transitioning to Length Measurer at the end of Grade 2 and continuing into Grade 3. In Grades 4 and 5, Abby's thinking was primarily at the Integrated Conceptual Path Measurer level.

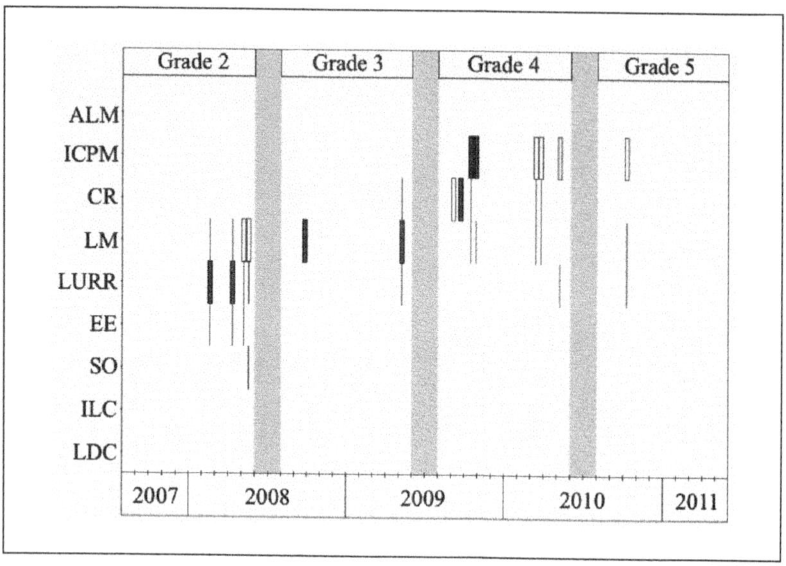

Figure 2.12. Abby's longitudinal growth chart. LDC = Length Direct Comparer, ILC = Length Indirect Comparer, SO = Serial Orderer to 6+, EE = End-to-End Length Measurer, LURR = Length Unit Relater and Repeater, LM = Length Measurer, CR = Conceptual Ruler Measurer, ICPM = Integrated Conceptual Path Measurer, and ALM = Abstract Length Measurer.

Anselm's growth. Anselm demonstrated behaviors consistent with thinking at the Length Unit Relater and Repeater level throughout Grade 2 and into Grade 3. During Grade 3, he made rapid progress through both the Length Measurer and Conceptual Ruler Measurer levels, arriving at the Integrated Conceptual Path Measurer level. During Grade 4, Anselm occasionally fell back to thinking at the Length Unit Relater and Repeater and Length Measurer levels on some tasks (see Figure 2.13).

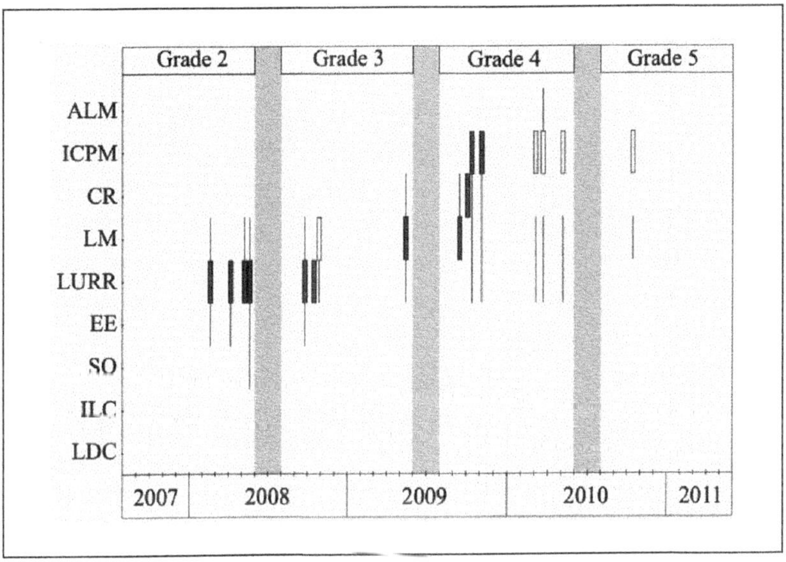

Figure 2.13. Anselm's longitudinal growth chart. LDC = Length Direct Comparer, ILC = Length Indirect Comparer, SO = Serial Orderer to 6+, EE = End-to-End Length Measurer, LURR = Length Unit Relater and Repeater, LM = Length Measurer, CR = Conceptual Ruler Measurer, ICPM = Integrated Conceptual Path Measurer, and ALM = Abstract Length Measurer.

Arielle's growth. Arielle did not show as much variability as Abby did during Grade 4. Instead, Arielle made steady progress from Length Unit Relater and Repeater to Length Measurer to Conceptual Ruler Measurer to Integrated Conceptual Path Measurer during the 4-year period. She spent approximately 1 year at each of the levels (see Figure 2.14).

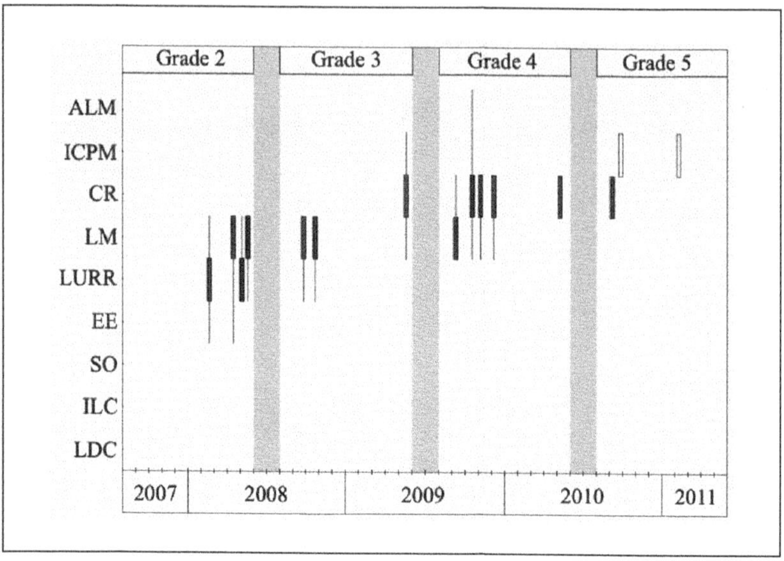

Figure 2.14. Arielle's longitudinal growth chart. LDC = Length Direct Comparer, ILC = Length Indirect Comparer, SO = Serial Orderer to 6+, EE = End-to-End Length Measurer, LURR = Length Unit Relater and Repeater, LM = Length Measurer, CR = Conceptual Ruler Measurer, ICPM = Integrated Conceptual Path Measurer, and ALM = Abstract Length Measurer.

Danny's growth. Danny demonstrated thinking at the Length Unit Relater and Repeater and Length Measurer levels during Grade 2 and Grade 3. During Grade 4, Danny demonstrated more variability in his thinking, illustrating thinking at the Length Measurer, Conceptual Ruler Measurer, and Integrated Conceptual Path Measurer levels. During Grade 5, Integrated Conceptual Path Measurer was his dominant level of thinking (see Figure 2.15).

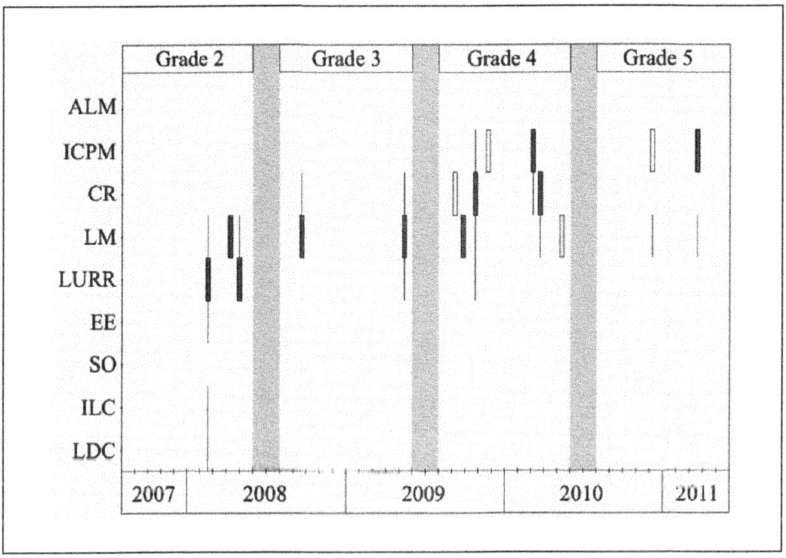

Figure 2.15. Danny's longitudinal growth chart. LDC = Length Direct Comparer, ILC = Length Indirect Comparer, SO = Serial Orderer to 6+, EE = End-to-End Length Measurer, LURR = Length Unit Relater and Repeater, LM = Length Measurer, CR = Conceptual Ruler Measurer, ICPM = Integrated Conceptual Path Measurer, and ALM = Abstract Length Measurer.

Drew's growth. Drew transitioned from Length Unit Relater and Repeater in Grade 2 to Conceptual Ruler Measurer in Grade 4 (see Figure 2.16). In only one teaching episode during Grade 3 did we identify his dominant level of thinking as Length Measurer. Although we observed behaviors consistent with thinking at the Abstract Length Measurer level during one teaching episode in Grade 4, no evidence of thinking at the Integrated Conceptual Path Measurer level occurred at that time, an apparent skip from Conceptual Ruler Measurer to Abstract Length Measurer. We could view this result as a potential disconfirmation of the sequence from Conceptual Ruler Measurer to Integrated Conceptual Path Measurer and into Abstract Length Measurer. Because we found only one such incidence, it raised our concern, but it did not motivate a change in the broader theoretical account of these levels. Further, researchers should expect to observe such skipping when observations are not continuous and rapid growth through levels is allowable within the theoretical. Drew did not participate in teaching episodes during Grade 5.

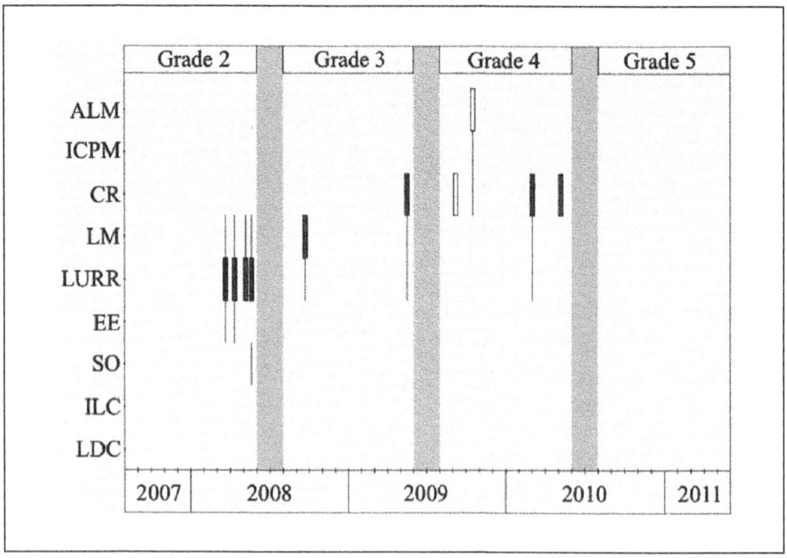

Figure 2.16. Drew's longitudinal growth chart. LDC = Length Direct Comparer, ILC = Length Indirect Comparer, SO = Serial Orderer to 6+, EE = End-to-End Length Measurer, LURR = Length Unit Relater and Repeater, LM = Length Measurer, CR = Conceptual Ruler Measurer, ICPM = Integrated Conceptual Path Measurer, and ALM = Abstract Length Measurer.

Owen's growth. Owen followed a pattern of approximately 1 year at each of three levels of thinking: Length Unit Relater and Repeater, Length Measurer, and Conceptual Ruler Measurer (see Figure 2.17). We did not observe any behaviors in Owen that were consistent with thinking at the highest two levels of the LT. Like Drew, Owen did not participate in teaching episodes during Grade 5.

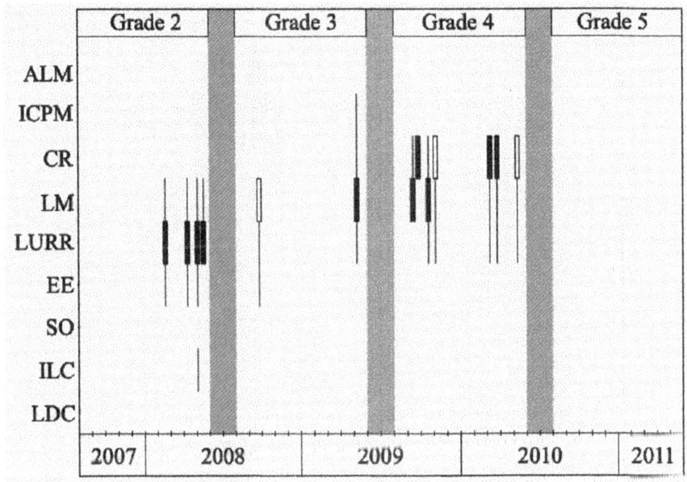

Figure 2.17. Owen's longitudinal growth chart. LDC = Length Direct Comparer, ILC = Length Indirect Comparer, SO = Serial Orderer to 6+, EE = End-to-End Length Measurer, LURR = Length Unit Relater and Repeater, LM = Length Measurer, CR = Conceptual Ruler Measurer, ICPM = Integrated Conceptual Path Measurer, and ALM = Abstract Length Measurer.

Composite growth chart for Grades 2 to 5. Several overall patterns emerged in the composite growth chart, shown in Figure 2.18. First, the dominant levels of thinking at the end of Grade 2 and the beginning of Grade 3 were Length Unit Relater and Repeater and Length Measurer. The Conceptual Ruler Measurer level appeared at the end of Grade 3 and continued into Grade 4. However, the variability of levels, both within each individual child and among the entire group of children, increased to encompass six levels (i.e., End-to-End Length Measurer,

Length Unit Relater and Repeater, Length Measurer, Conceptual Ruler Measurer, Integrated Conceptual Path Measurer, and Abstract Length Measurer). During Grade 5, the variability decreased somewhat, with Integrated Conceptual Path Measurer most often emerging as the dominant level.

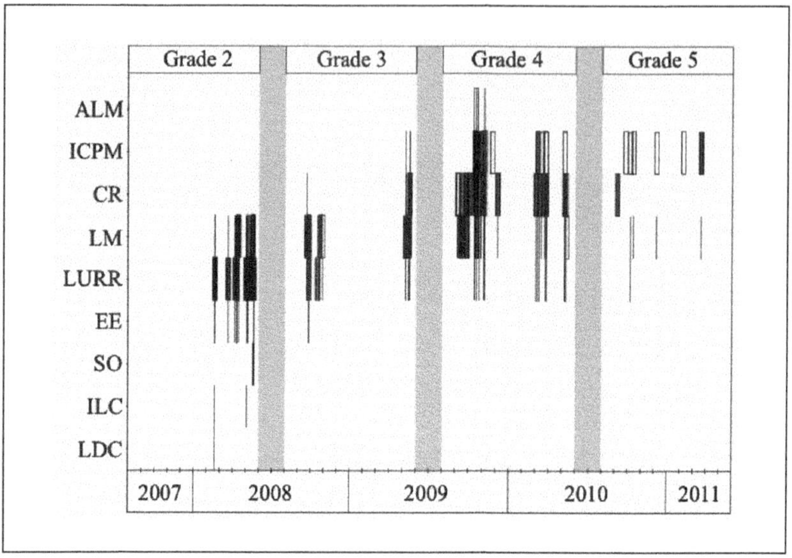

Figure 2.18. Grades 2–5 composite growth chart. LDC = Length Direct Comparer, ILC = Length Indirect Comparer, SO = Serial Orderer to 6+, EE = End-to-End Length Measurer, LURR = Length Unit Relater and Repeater, LM = Length Measurer, CR = Conceptual Ruler Measurer, ICPM = Integrated Conceptual Path Measurer, and ALM = Abstract Length Measurer.

When comparing individual growth charts with the composite growth chart, we noticed that Anselm's growth accelerated rapidly (see Figure 2.13), but it was later than that of the other children (see Figure 2.18). Indeed, his overall gain across 4 years was at least as pronounced as that of others in this cohort. His case, as shown by the shape of this developmental change, illustrates the potential for children to make rapid corrections toward a more deliberate incremental growth pattern.

THE REVISED DEVELOPMENTAL PROGRESSION OF THE LEARNING TRAJECTORY FOR LENGTH MEASUREMENT

Results from the qualitative and quantitative data analyses outlined in the previous sections confirmed that, overall, the developmental progression of the initial LT for length measurement (Sarama & Clements, 2009b) served as an appropriate tool for designing length comparison and measurement tasks, predicting and analyzing children's responses, and identifying shifts in thinking. However, these findings also suggested the need to refine the developmental progression by collapsing and revising some levels of the initial developmental progression, as well as elaborating the developmental progression by articulating more sophisticated levels. Findings from the study also provided insights into foundational schemes that begin to develop within a level that serve as psychological foundations for growth into subsequent levels. The revised developmental progression includes these notes on psychological foundations. For each level, we provide evidence that supports modifications to the levels of the initial developmental progression and conclude with the presentation of a revised developmental progression for the LT for length measurement.

Pre-Length Quantity Recognizer

None of the participants in the Northeast cohort or the Midwest cohort showed evidence of thinking at the Pre-Length Quantity Recognizer level of the LT for length measurement. This outcome suggests that children are able to recognize length as an attribute by pre-K. Furthermore, our research team observed that this pre-length level is not a productive instructional objective for teaching and learning length measurement (i.e., we want to avoid the possible misinterpretation that curriculum writers and teachers should attend to every level—including this one). Therefore, the revised developmental progression of the LT for length measurement omitted this level.

Length Quantity Recognizer

Our research team found that children in the pre-K to Grade 2 cohort intermittently demonstrated thinking at the Length Quantity Recognizer level throughout the 4 years of the study. Results from the Rasch analysis indicate that Length Quantity Recognizer is a level of the LT that develops early; however, the Length Quantity Recognizer items appeared to be more difficult than the Length Direct Comparer items. We attribute this inconsistency between the qualitative data from the teaching experiment and quantitative data from the Rasch analysis to the difficulty inherent in designing an assessment item that can determine whether a child recognizes length as an attribute without asking a comparison identification of linear extent in an explicit way. Therefore, despite this inconsistency between the qualitative and quantitative data relevant to the Length Quantity Recognizer level, we regarded this level as confirmed in the study.

Length Direct Comparer

The children in the study used the Length Direct Comparer level predominantly throughout their pre-K and kindergarten years and intermittently in Grades 1 and 2. Results from the longitudinal teaching experiment confirmed that children at this level compare objects by physically or mentally sliding, rotating, or flipping them to align them at one endpoint. Although, as previously mentioned, the Rasch analysis indicated that assessment items designed for the Length Quantity Recognizer level were more difficult than items designed to assess this level, our research team did not regard this inconsistency within the qualitative and quantitative data as disconfirming evidence.

Length Indirect Comparer

Our findings indicated that the Length Indirect Comparer (previously Indirect Length Comparer) level required modification. Children in the study showed evidence of thinking at the Length Indirect Comparer level of the LT throughout pre-K through Grade 2. However, children often began to use explicit transitive reasoning in length-measurement contexts at the same time that they began demonstrating behaviors consistent with the subsequent measurement levels. Findings from the Rasch analysis showed that the Length Indirect Comparer, Serial Orderer to 6+, and End-to-End Length Measurer levels do not occur independently of one another. Therefore, in the revised LT, indirect length comparison develops idiosyncratically and in parallel with the subsequent measurement levels.

This observation is important because the placement of direct and indirect comparison has varied, and we believe that our evidence helps resolve this debate. That is, in our initial LT (Sarama & Clements, 2009b), we placed Indirect Length Comparer where the extant literature indicated, after Length Direct Comparer and before End-to-End Length Measurer. Other models considered direct and indirect comparison to constitute a separate nonmeasurement developmental progression (Barrett & Battista, 2014). Our data show a complex relationship between the two. First, direct comparison is cognitively foundational for other nonnumerical processes (i.e., indirect comparison) as well as for the numerical, or measurement, processes. That is, the actions on objects of the Length Direct Comparer level (such as alignment and comparison of the endpoints of lengths) related hierarchically to all subsequent length development. Further, indirect comparison develops in tandem with the development of more sophisticated measurement processes. Therefore, the evidence supports a third model. Nonnumerical comparison does not develop before measurement, nor is it a separate LT. (We believe that the latter view separates them on mathematical, not psychological, bases.) More accurately, we see a single LT for length measurement, with two subtrajectories (numerical and nonnumerical length measurement), interacting but only loosely coupled, with growth in each supporting growth in the other.

Serial Orderer to 6+: Incorporated Into Length Indirect Comparer

Results from the initial assessment administered to the Northeast cohort indicate that the children in pre-K were not yet able to coordinate multiple length comparisons to establish an ordered sequence of objects. By the end of pre-K, however, the research team observed that children were able to use spatial cues to coordinate multiple comparisons of linear extent and create an ordered sequence of objects. These children also showed evidence that they had begun to develop the implicit concept that lengths can consist of repetitions of shorter lengths laid end to end, which we had originally hypothesized to develop at the subsequent level. As described in the previous Length Indirect Comparer section, results from the Rasch analysis indicated that the Length Indirect Comparer, Serial Orderer to 6+, and End-to-End Length Measurer levels of the LT occur simultaneously. For those reasons, the research team collapsed the Serial Orderer to 6+ level into the Length Indirect Comparer level of the LT for length measurement.

End-to-End Length Measurer

Results indicate that children showed evidence of laying units end to end to create measures for comparison, which is indicative of thinking at the End-to-End Length Measurer level from kindergarten through Grade 2. Rasch analysis indicated that this level occurred alongside Length Indirect Comparer and the Serial Orderer to 6+ levels but before the Length Unit Relater and Repeater level. These findings lend broad support to this level. Thus, we considered this level validated and sequenced between the Length Direct Comparer and Length Unit Relater and Repeater levels (with the new Length Indirect Comparer level developing in parallel; see Table 2.1).

Table 2.1
A Comparison of the Levels From the Initial and the Revised Developmental Progressions for Length Measurement

Initial Developmental Progression		Revised Developmental Progression
Pre-Length Quantity Recognizer		
Length Quantity Recognizer	→	Length Quantity Recognizer
Length Direct Comparer	→	Length Direct Comparer
Length Indirect Comparer Serial Orderer to 6+	}→	Length Indirect Comparer
End-to-End Length Measurer	→	End-to-End Length Measurer
Length Unit Relater and Repeater	→	Length Unit Relater and Repeater
Length Measurer	→	Consistent Length Measurer
Conceptual Ruler Measurer	→	Conceptual Ruler Measurer Integrated Conceptual Path Measurer Abstract Length Measurer

Length Unit Relater and Repeater

As early as kindergarten, children in the Northeast cohort showed evidence of iterating a single unit to measure, that is, using a Length Unit Relater and Repeater approach for measuring. However, children in the Midwest cohort predominantly used thinking at the Length Unit Relater and Repeater level in Grade 2 and intermittently across Grades 3 to 5. Results from the Rasch analysis indicated that the Length Unit Relater and Repeater level extended from the upper reaches of the previous level and in front of the subsequent level of the LT for length measurement. Therefore, qualitative and quantitative findings confirmed this level.

Length Measurer

We changed the name of this level to *Consistent Length Measurer* to characterize changes in the quality of children's behaviors and thinking. We observed thinking at the level in all eight children at the Midwest site, but primarily in Grades 3 and 4. Our findings confirmed this level of the LT.

Conceptual Ruler Measurer

Children in Grade 4 exhibited extensive use of this thinking. At least one child exhibited this thinking pattern in Grade 3, and another child continued using it through Grade 5. Our findings confirmed this level of the LT. Although the Rasch analysis showed some overlap between Conceptual Ruler Measurer and Consistent Length Measurer, the composite growth chart in Figure 2.18 indicated clear distinctions in the developmental timing of the two levels.

Conceptual Ruler Measurer thinking subsumes and extends the schemes exhibited at the Length Unit Relater and Repeater level because children at that level can iterate units that they can represent internally without needing perceptual support for the unit parts or the partitioning of the object. The important schemes include mental iterating, partitioning, or unitizing schemes, working in tandem with a coordinating number and space scheme. Children operating at this level can project a unit onto a linear space by imagining the repeated collection of a unit image, or they can partition the object by imagining the subdivision markers along the span of the object. A caveat is that this level of thinking does not yet involve integration for multiple units and units of units along a bent path. Thus, children are likely to produce imprecise or inaccurate measures of bent paths.

Integrated Conceptual Path Measurer

As previously noted, our initial LT for length measurement presented a developmental progression only through the primary grades. Because our study included older children, our findings suggested the need for expanding the initial LT for length measurement by adding the Integrated Conceptual Path Measurer level.

Our quantitative analysis distinguished Integrated Conceptual Path Measurer from Conceptual Ruler Measurer; and the composite growth chart, shown in Figure 2.18, also furnishes evidence of this distinction. The Integrated Conceptual Path Measurer level was evident among all six focus children in the Midwest cohort during Grades 4 and 5, but it was more prevalent with four of them; it was more prevalent for four of the six children in Grade 5 than the Conceptual Ruler Measurer level.

The Integrated Conceptual Path Measurer level of thinking extends and reorganizes the Conceptual Ruler Measurer level of thinking. At that level, a child attends to multiple aspects and subcollections of units across the parts of complex bent paths or along the perimeter of a polygon. The child may enact a multiplicative comparison scheme for units and for collections of units of length across several related cases of a given object (e.g., moving a bend in a path from one portion of the path to another). A child thinking at the Integrated Conceptual Path Measurer level notices the relevant quantities in complex geometric situations and finds ways of preserving the overall length of a path while changing its shape by compensating among the parts of that path. This kind of operation is possible at the Integrated Conceptual Path Measurer level because of the enactment of further coordination among units, subunits, and even superunits. Still, the child may not necessarily be able to integrate a large set of polygons by perimeter because of continuing reliance on immediate cases of figures. However, at the Abstract Length Measurer level, the child's scheme for integrating space and number extends to include angle, time, and a broader nesting of units, superunits, and subunits simultaneously.

Abstract Length Measurer

Three of the six children in the Midwest cohort study exhibited thinking beyond the previously described levels. Findings from our previous studies pointed to the existence of children's dynamic and integrated reasoning among collections of related geometric figures to analyze the underlying relationship (Barrett et al., 2006). During the children's fourth- and fifth-grade years, we saw glimpses of thinking at the Abstract Length Measurer level, but we would expect older children to exhibit this level of thinking on a more consistent basis. These findings resulted in our modifying the LT by adding this more sophisticated level.

Thinking at the Abstract Length Measurer level includes an advance over thinking at the Integrated Conceptual Path Measurer level when the child exhibits a reflective capacity to organize sequences of measurement activity among complex, and even curved, paths with attention to a broader multiplicative comparison to set up ratios or to develop coordination with other measures such as time or angle. These measures can be developments of derived units that integrate length within a larger investigation. Figure 2.19 shows the revised developmental progression for the LT for length measurement.

Developmental Progression	Mental Actions on Objects	Anticipated Misconceptions or Partial Conceptions
Length Quantity Recognizer (LQR) Identifies length and distance as attributes.	Initially, action schemes, both physical/kinesthetic and visual, implicitly trace linear extents. In certain situations, children are sensitive to linear extent from infancy, including comparisons and accumulations (one object is longer if a scan lasts perceptibly longer than the scan of another object). Later, connects action schemes with length vocabulary. Connects vocabulary with categories of linear extent, such as *tall/long* or *short*. *Notes on psychological foundations: Children make intuitive comparisons on a direct perceptual, normative basis (one object can be a class standard, scored in memory, such as a doll's length), or a functional basis (if guided or prompted; e.g., is this block long enough?).*	May compare size generally without particular dimensions (confounds length, area, and volume). *"This is long. Everything straight is long. If it's not straight, it can't be long."* May attend to salient differences at one end of the objects rather than a scan (potentially leading to inaccuracies if the other endpoints do not align). Irrelevant details, such as shape, may affect these categorizations and comparisons.
Length Comparer (LC) ***Length Direct Comparer (LDC)*** Physically aligns two objects to determine which is longer or whether they are the same length. *Stands two sticks up next to each other on a table and says, "This one's bigger."* May use a ruler (as a stick rather than as a measuring tool) to directly compare it and another object. Uses the following terms: *long, longer, longest*.	The action scheme addresses length as a linear extent from endpoint to endpoint of a path. Shape of the objects and path may affect the application of the scheme. With perceptual support, can physically or mentally compare objects. For example, can physically slide, rotate, or flip concrete objects to align and compare the endpoints. Eventually, internalizes these actions, generalizing the action scheme to allow for mental comparison of lengths of objects.	When the researcher asks a child to compare objects that are not adjacent and cannot be physically aligned, the child may say, "they are too far away." May assign numbers to lengths of objects without being consistent with perceptual comparisons (e.g., a child may assign a greater number to the object that he or she found to be shorter).

Figure 2.19 continues on next page

Developmental Progression	Mental Actions on Objects	Anticipated Misconceptions or Partial Conceptions
Length Indirect Comparer (LIC) Note: Nonnumerical comparison is a subtrajectory within this LT. This level begins to develop in parallel with the subsequent measurement levels over time. Compares the length of two objects by representing them with a third object. *Uses a piece of string to compare the lengths of two objects.* Uses the following terms: *long, longer, longest, short, shorter, shortest.*	Applies explicit transitive reasoning to the linear extent of objects or representations of objects, allowing for the comparison of two objects mediated by a third. Begins coordinating multiple length comparisons to establish an ordered sequence of objects and uses spatial cues to represent order.	When the child compares two objects by using a third object that is longer than both, may not be able to use part of the longer object as a referent for comparing the shorter objects.
End-to-End Length Measurer (EE) Lays units end to end to create measures for comparison. *When asked, "How many light-colored strips long is the dark-colored paper strip?" the child succeeds, ignoring extra light-colored strips (beginning and end).*  Although can produce a meaningful measure when given all the objects, may use fewer objects or a single object to measure. Uses rulers with minimal guidance and some success.	A counting scheme operates on an intuitive unit of spatial extent or amount of movement, directing physical (or, less frequently, eye movements) along a length while counting (resulting in a trace-and-count or point-and-count strategy). This action scheme develops into an implicit understanding that the child can compose a linear extent by repetition of shorter lengths. Increased attention to composing a length from parts or sweeping through a length with parts enhances an initial weak intuitive constraint to use equal-size units or to avoid gaps between "units."	May not recognize the need for equal-length units. When provided with fewer unit pieces than needed to span a length, may assign a numerical value by moving along the length while counting. When asked to measure a line segment, moves finger along segment, saying, 1, 2, 3, 4, 5, without relating motions to units. May be able to measure with a ruler but often lacks understanding or adaptive skills (e.g., ignores zero point or reports the number at the far end of the ruler).

Figure 2.19 continues on next page

Developmental Progression	Mental Actions on Objects	Anticipated Misconceptions or Partial Conceptions
Length Unit Relater and Repeater (LURR) Iterates a single unit to measure. Relates size and number of units, at least intuitively. *"If you measure with centimeters, not inches, you'll need more because each one is smaller."* Adds two lengths to obtain the length of the whole. Uses rulers with minimal guidance and more success than at the previous level. Measures a length accurately with a ruler if aligned to the zero point.	Curtails the previous level's fill-and-count process (e.g., by pointing and sweeping in a coordinated set of actions), leading to the development of a new action scheme that includes the ability to iterate a unit along a perceptually available object. Can maintain the image of each placement while moving the physical unit to the next iterative position (initially with weak constraints on this placement) and counting. With the support of a perceptual context, can instantiate the scheme with large and small units to determine that fewer of the large units would span an object. *Notes on psychological foundations: Action schemes allow the application of counting-all addition schemes to apply to measures, which allows additive reasoning to compare length measures by using whole units.* A partitioning scheme provides linkages from partial collections of iterated unit images to the entire collection.	Measures by repeated use of a unit, imprecise initially in broken-ruler tasks and without a zero point for reference. Iterates a single unit, sometimes with mistakes, to find how many would stretch end to end. Can start from 5 and iterate one or more light-colored strips to find the length of 8 strips.  When an object does not align to start at zero on a ruler, children may include the initial tick mark when they count tick marks to find length. May measure length by aligning the object to the numeral 1 on the ruler.

Figure 2.19 continues on next page

Developmental Progression	Mental Actions on Objects	Anticipated Misconceptions or Partial Conceptions
Consistent Length Measurer (CLM) Measures straight paths consistently, knowing the need for identical units, relationship between different units, partitions of unit, zero point on rulers, and accumulation of distance. Can resolve broken-ruler tasks effectively. Begins to relate and coordinate units and subunits. For example, may use both meters and centimeters when measuring. *While finding the length of a 3.14-meter-long rope, says, "it's 3 meters, 14 centimeters in all."* Compares two bent paths by relating the parts in a 1:1 relation of successive parts. Begins to estimate reasonably. Iterates a single length unit without gaps or overlaps along a straight path.	Use of rulers integrates intervals and the marks that indicate endpoints for those intervals to establish linear quantity. Identifies zero as the starting point for motion along a ruler and the numbers as reports of completed motion. The integration of counting and iterating schemes allows children to measure line segments with physical units or rulers by concurrently iterating a unit and subdividing a unit to increase precision. May identify that each unit is countable as if it were an indivisible stick; thus, understands that a length is a ratio comparison between the unit object and the measured object. *Notes on psychological foundations: The length scheme has additional hierarchical components, including the ability to conceive of an object's length as both a total extent and a comparison of units. This scheme adds constraints on the use of equal-length units. Recognizes that partitioning units themselves is possible to increase precision, allowing the accurate use of units and subordinate units. Although the scheme allows for multiplicative comparisons in simple situations, production of additive comparisons is more likely.*	May make errors when considering the length of a bent path as the sum of its parts. For example: • May not equate the sum of the parts of a bent path to the length of the whole because of an avoidable accumulation of rounding errors. • May not be perturbed by geometric inconsistencies. • May misidentify the attribute or identify the wrong attribute when measuring perimeters or complex paths. Numeric answers for the perimeter may include a count of both area and length units. May impose units on a bent path but miscount at corners.

Figure 2.19 continues on next page

Developmental Progression	Mental Actions on Objects	Anticipated Misconceptions or Partial Conceptions
Conceptual Ruler Measurer (CRM) Has an "internal" measurement tool. Mentally moves along an object, segmenting it and counting the segments. Operates arithmetically on measures. Projects or translates given lengths to determine missing lengths. When finding the length of the path below, uses the vertical length of the first leg to determine the combined vertical length of third, fifth, and seventh legs of the path. 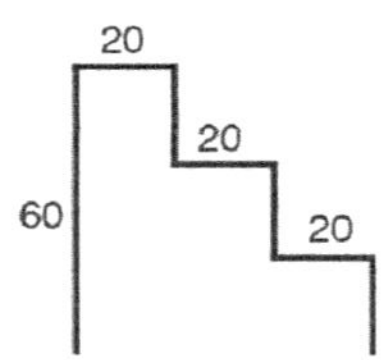 Notices or is perturbed by geometric inconsistencies. Estimates the length of an object that is not partitioned with accuracy and estimates without any available image of the standard unit. Employs explicit strategies to estimate lengths, including developing benchmarks for units (e.g., a 1-in. piece of gum) and composite units (e.g., a 6-in. dollar bill) and mentally iterating those units.	The length scheme allows mental partitioning of a length into a given number of equal-length parts or the mental estimation of length by projecting an imagined unit, ruler, or a collected sequence of larger units onto unpartitioned objects, even without perceptually available markers to indicate partitioning of the object. The schemes of this level include mental iterating, partitioning, or unitizing schemes working in tandem with a coordinating number and space scheme. *Note on psychological foundations: By using these schemes, children increasingly employ multiplicative reasoning in comparison cases.*	Still measures straight paths consistently but may make errors with bent paths. Is not yet able to draw and coordinate a comprehensive set of cases as evidence of an underlying pattern. May not coordinate lengths among multiple bent paths.

Figure 2.19 continues on next page

Developmental Progression	Mental Actions on Objects	Anticipated Misconceptions or Partial Conceptions
Integrated Conceptual Path Measurer (ICPM) Computes perimeter of a polygon, including complex cases. Finds several related cases of polygons with the same perimeter or the same area and relates those cases to one another by logical comparison to provide evidence of an underlying pattern. Rearranges portions of a path A to B without number measures for the parts involved, arguing that both paths are the same length on the basis of geometric properties. Makes changes in one part of a figure and adjusts other sides to compensate for changes of length to maintain the fixed overall path length. Makes claims about sets of complex paths on the basis of perimeter and other linear aspects of the objects. Can analyze length within two- and three-dimensional contexts. In selection of units, shows well-developed ideas of precision and accuracy.	Integrates and compares sets of units along each section of a given path. Operates on bent paths by partitioning to establish units, treating a group of units as a flexible object. This object is a "string" of units wrapped around the entire perimeter or along the entire path. Anticipates and monitors sets of related cases of a figure. Coordinates related measures when they "rezero" the ruler to measure complex paths. Operates efficiently and precisely with subunits, units, and units of units. Begins to coordinate other measures with linear measures, such as angle, curvature, or time. Extends the length scheme to incorporate multiple units and collections of units to compare related paths or perimeters from sets of polygons. *Notes on psychological foundations: This improved scheme integrates sets of units along each section of a path or object.*	May not coordinate a series of changes in a systematic way across multiple figures to formulate and justify a valid argument, despite the ability to compensate within a single figure.

Figure 2.19 continues on next page

Developmental Progression	Mental Actions on Objects	Anticipated Misconceptions or Partial Conceptions
Abstract Length Measurer (ALM) Organizes and synthesizes sets of objects on the basis of perimeter or collections of complex bent paths on the basis of overall length in two- or three-dimensional contexts to formulate and justify a valid argument. Constructs derived units with linear measures, such as miles per hour, and makes appropriate unit conversions with derived measures. Computes perimeter or path length, including units and divisions of units including measures of nonintegral values. Can explain that this subdivision process is potentially unlimited. Notices and is perturbed by geometric inconsistencies within polygons. *When the researcher asks a child to check the validity of a hinged model for a triangle "case" with sides of 5, 7, and 12, the child can provide a rationale for excluding such cases by associating angle size and side length by showing that for the sides to meet away from the base side (12), they must be stretched out, yielding a sum of something more than 12 (and they must be increasingly longer as the angle of the sides increases).* Uses measures as evidence of the existence of patterns and relationships between quantities and constructs models that describe these patterns and relationships (e.g., can use lower and upper bounds of linear estimates to order several curves by length). Measures to the degree of precision allowed by a tool by estimating to a fraction of the smallest calibration mark provided on the instrument.	Engages in proportional reasoning about repeated or coordinated cases of paths to reflect on patterns in these cases. Coordinates multiplicative and additive reasoning in fluent ways. Integrates multiple collections of units across entire sets of paths or objects. Develops a continuous sense of length. *Notes on psychological foundations: This integrated action scheme engages dynamic imagery to coordinate and operate internally on collections of units of units and on collections of complex paths. The scheme allows children to argue about derived units as another attribute.*	

Figure 2.19. Revised developmental progression component of the LT for length measurement.

SECTION II

Chapter 3

Area

Douglas H. Clements
University of Denver
Julie Sarama
University of Denver
Amanda L. Miller
Illinois State University

As with length, the measurement of area connects with other mathematical topics and with children's experiences with the physical world. Developing concepts and skills in area measurement builds a foundation for understanding topics from multiplication to fractions, as well as for solving real-world problems ranging from cartography, deforestation, or navigation to the designing of packing material or the measuring of molecules on a surface with a nanomeasuring machine.

We define area as the amount of two-dimensional space that a planar closed region contains. Some assumptions of area measurement are that we choose a two-dimensional region as a unit, congruent regions have equal areas, regions do not overlap, and the area of the union of two nonoverlapping regions is the sum of their areas (cf. Lehrer, 2003; Lehrer, Jacobson, et al., 1998; Sarama & Clements, 2009b). Thus, finding the area of a region involves covering (or subdividing) a region with a two-dimensional unit of measure.

Despite attention to area measurement in and out of schools, research has demonstrated that children often perform rote procedures without understanding several crucial area concepts and principles. For more than three decades, children in the United States have performed poorly on measurement tasks, including area measurement. Only 51% and 48% of 13-year olds correctly answered that the area of a 6 cm by 4 cm rectangle was 24 square centimeters on the Second (1978) and Third (1982) National Assessment of Educational Progress (NAEP) exams, respectively (National Assessment of Educational Progress [NAEP], 1983). Results deteriorated on a similar item on the Fourth (1986) NAEP exam; only 46% of seventh graders correctly determined the area of a 5 by 6 rectangle (Lindquist & Kouba, 1989). At the present time, children still struggle with area-measurement tasks. On the 2011 NAEP, only 24% of fourth graders were able to choose the correct value for the area of a square, and measurement was the weakest of five substrands for fourth and eighth graders (NAEP, 2013).

In investigating why children struggle, researchers have analyzed textbooks and manipulative materials (J. P. Smith, Males, Dietiker, Lee, & Mosier, 2013), as well as children's common errors and misconceptions. One of these misconceptions stems from measuring a continuous quantity with discrete objects (e.g., foam squares). Cavanaugh (2008) found that "Many textbooks present regions which are already partitioned so that students need only count the squares one by one to discover the area of a shape" (p. 55). Such representations encourage children to curtail the learning of area conceptions to rote counting of preformed units or the use of formulas without adequate constraints or adjustments to context. This curtailment caused some children to develop an incomplete understanding of fundamental area concepts, including understanding the attribute of area, unit concepts, accumulation and additivity, conservation, and the relationship between number and space (Lehrer, 2003; Lehrer, Jacobson, et al., 1998; Sarama & Clements, 2009b).

SCHEMES SUPPORTING THE DEVELOPMENT OF AREA MEASUREMENT

These results indicate that measuring area is more than a procedure or skill that children can quickly learn and retain. Instead, it requires the development and coordination of multiple concepts, strategies, and schemes (cf. Kordaki & Potari, 2002; Outhred & Mitchelmore, 2000; Sarama & Clements, 2009b). We next discuss four schemes that support the development of area measurement: covering, quantifying, subdividing, and spatial structuring. These schemes were present in the initial developmental progression of the learning trajectory (LT) for area measurement (Sarama & Clements, 2009b) and serve as a starting point for our research reported here.

Covering

In an area-measurement context with rectilinear regions, covering may refer to at least two different activities: tiling or iterating. Young children may cover by tiling with multiple copies of a unit tile (cf. End-to-End Length Measurer level of the LT for length measurement). When a researcher provides a rectilinear region and a collection of tiles (as long as the number of tiles in the collection is either greater than or equal to the number of unit tiles necessary to cover the region), the child may use unit tiles to cover the two-dimensional space within the region, with varying degrees of success. Older children may reason about covering by iterating—placing unit tiles end to end (cf. Length Unit Relater and Repeater level of the LT for length measurement). If the researcher presents the child with a rectilinear region and fewer than the number of tiles necessary to cover it, the child may translate one or more tiles successively to imagine a covering of the region, again with varying degrees of success. For these two related schemes of tiling and iterating (in their fully realized forms), children recognize that the situation requires that two-dimensional units cover a region, that both schemes instantiate a process of covering, and that the children may produce a complete covering without gaps or overlaps. The type of operation enacted differentiates the schemes: With the tiling scheme, the fit of the tiles themselves in a shape that matches a given outline constrains using physical movement; and with the iterating scheme, iterating small sets of physical tiles requires more internal organization of the movements of some small set of tiles or some kind of drawing of contiguous unit tile shapes. In both cases, while the child learns about space-covering and space-filling notions, the child begins to appreciate the consequences of leaving gaps or overlaps (Lehrer, 2003) and extending over the boundary (Lehrer, Jenkins, & Osana, 1998); and the schemes incorporate those constraints.

Quantifying

Quantifying is a second scheme related to area measurement. It refers to how a child determines the number of items in a collection. In an area-measurement task, quantification is usually the enumeration of square units that cover a region. Young children have a tendency to count by ones, either systematically or unsystematically. Because young children with prenumerical counting schemes have not yet coordinated their pointing acts with their number-word sequence, they may double-count squares, omit others, and seem untroubled by arriving at different "counts" (Olive, 2001). Eventually, these children learn to organize their counting activity by marking or labeling objects while they count. Regardless, children with prenumerical counting schemes not only need to count each of the objects in a collection, but they also need to have the objects perceptually available for counting. When some or all of the objects are not perceptually available, children may count around the border, spiraling to the center. Other children count left to right, up and down, or in S-like patterns. Children with more sophisticated quantifying schemes group objects together. They may skip-count or repeatedly add the number in a group. In time, children develop an understanding of grouping and place value and connect repeated addition with multiplication. Older children have a fully multiplicative enumeration scheme, supporting a conceptual understanding of Cartesian multiplication and its applications.

Subdividing

Subdividing schemes are also relevant to area measurement. Subdividing is the mental operation of breaking a continuous whole into countable parts. Because young children have not yet anticipated the relationship between the parts or the relationship between the parts and the whole, they may be content with unequal-sized parts or pieces. These children may not recognize that the subdivided whole is equal to the sum of its parts (Piaget, Inhelder, & Szeminska, 1960). Children with more geometrically sophisticated subdividing schemes anticipate these relationships. For example, when a researcher presents children with a rectangular region, children may devise subdividing schemes that use the length of the sides of the rectangle to determine where to draw parallel row and column line segments to create identical square units. They have developed an understanding of the relationship between the whole, the equal-sized parts, and the number of these parts.

Spatial Structuring

Much of the recent research on area measurement has focused on the competence of a fourth scheme, spatial structuring. Spatial structuring is the mental operation of constructing an organization or form for an object or set of objects in space (Sarama & Clements, 2009b). The origins of this construct lie in work by Piaget (e.g., Piaget & Inhelder, 1948/1967) and observations and analyses by Outhred and Mitchelmore (1992). Spatial structuring is a form of abstraction—the process of selecting, coordinating, unifying, and registering in memory a set of mental objects and actions. Research indicates that such spatial structuring precedes meaningful mathematical use of the structures, such as determining area or volume (Battista & Clements, 1996; Battista, Clements, Arnoff, Battista, & Borrow, 1998; Outhred & Mitchelmore, 1992). For example, unless a child can conceptualize a rectangular region as consisting of unit squares, unitized into a row (or column), with that row iterated a number of times equal to the number of squares in each column (or row), that child often will use the formula

$$\text{area} = \text{length} \times \text{width}$$

only as a rote procedure, without understanding its mathematical foundation. The scheme of spatial structuring in this context involves recognizing the goal of partitioning a region into parts and the activity of organizing the region into a row-and-column structure, with the result of a fully partitioned and quantifiable region (ultimately the structure is related to the linear dimensions of the rectangular region).

A LEARNING TRAJECTORY FOR AREA MEASUREMENT

Research—on these schemes and others—indicates that understanding area involves a developmental progress that begins with a primitive spatial notion that the region must be covered, with no gaps or overlaps. Young children may attempt to match only corresponding sides of shapes to compare them (e.g., when judging congruence; Beilin, 1982). Although knowing whether children are conceptualizing two-dimensional shapes (congruence or area) or just the one-dimensional side (length) is difficult, these studies and our informal experiences suggest that children may believe that they are comparing shapes by using this side-matching strategy. A study by Lehrer, Jenkins, and Osana (1998) indicated that 73% of primary-grade children did not display this understanding and accepted circles placed within a square region as a viable way to measure area. They appeared to use boundedness rather than space filling to judge the adequacy of covering (Lehrer, Jenkins, et al., 1998). After they achieve this conceptual base, children learn increasingly sophisticated levels of spatial structuring. This research corpus, along with other research on area measurement, formed the basis for the initial LT for area measurement. Figure 3.1 presents the developmental progression along with the mental actions on objects that the original authors hypothesized underlie the behaviors indicative of each level of thinking (Sarama & Clements, 2009b).

This developmental progression of the LT for area measurement was the starting point for our research. However, from the initial research review (Clements & Sarama, 2007; Sarama & Clements, 2009b) to the beginning of the longitudinal study reported in this monograph, we were acutely aware of the limitations of the research base. Overall, only a limited number of studies discussed children's learning of area. Further, the notion of spatial structuring tended to dominate many of these studies (e.g., Battista & Clements, 1996; Battista et al., 1998; Outhred & Mitchelmore, 1992). We believe that this notion is a critical component of understanding area, but we also believe that other components exist, many of which have received little attention from researchers. The three chapters that follow describe our efforts to test, refine, and extend the initial developmental progression of the LT for area measurement presented in Figure 3.1.

Developmental Progression	Actions on Objects
Pre-Area Quantity Recognizer. Shows little specific concept of area. Uses side-matching strategies in comparing areas. May draw approximation of circles or other figures in a rectangular tiling task.	Perceives space and objects within the space.
Area Simple Comparer. May compare areas by using only one side of figures or by estimating on the basis of length plus (not times) width. May compare areas if task suggests superposition or unit iteration. Measures area with ruler, measuring a length, then moving the ruler and measuring that length again, apparently treating length as a two-dimensional space-filling attribute.	By using perceptual objects, internal innate or early developing competencies, compares the extent of regions (intuitively, perceptually).

Figure 3.1 continues on next page

Developmental Progression	Actions on Objects
Side-to-Side Area Measurer. Covers a rectangular space with physical tiles. Children have yet to organize, coordinate, and structure two-dimensional space, however, without such perceptual support. In drawing (or imagining and pointing to count), can represent only certain aspects of that structure, such as approximately rectangular shapes next to one another. Covers a region with physical tiles and counts them by removing them one by one. Draws within the region in an attempt to cover the region. May fill only next to existing guides (e.g., sides of region). 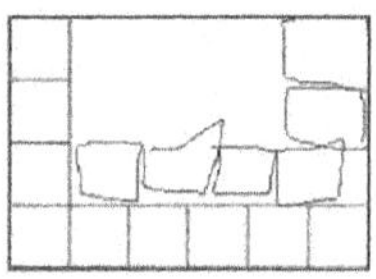	With perceptual support, can visualize that regions can cover regions. With strong guidance and perceptual support from materials, can direct the covering of that space and recognize that covering as complete. May represent concatenations of rectangular shapes, aligning them, but intuitively and in one dimension.
Primitive Coverer. Draws a complete covering, but with some errors of alignment. Counts drawn shapes around the border, then unsystematically in the interiors, counting some twice and skipping others. 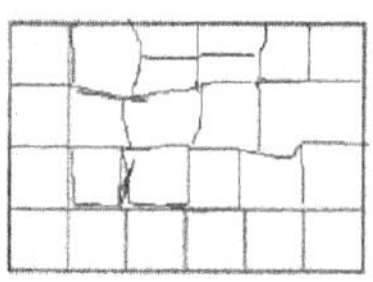	Applies explicit understanding that must cover entire region with shapes. Implicit visual patterning of concatenations guides placement of squares in rows.

Figure 3.1 continues on next page

Developmental Progression	Actions on Objects
Area Unit Relater and Repeater. Draws as above. Also, counts correctly, aided by counting one row at a time and often by perceptual labeling.	Stronger constraints on object counting and use of rows as an intuitive structure or explicit application of labeling as marker allows child to keep track.
Partial Row Structurer. Draws and counts some but not all rows. May make several rows and then revert to making individual squares, but aligns them in columns. Does not coordinate the width and height. In measurement contexts, does not necessarily use the dimensions of the rectangle to constrain the unit size.	Builds, maintains, and manipulates mental images of composite shapes, structuring them as composites of individual shapes and as a single entity—a row. Applies this composite unit repeatedly, but not necessarily exhaustively.
Area Row and Column Structurer. Draws and counts rows as rows, drawing with parallel lines. Counts the number of squares by iterating the number in each row, by using either physical objects or an estimate for the number of times to iterate. Those who count by ones usually do so with a systematic spatial strategy (e.g., by row). If the task is to measure an unmarked rectangular region, measures one dimension to determine the size of the iterated squares and eventually measures both, to determine the number of rows needed in drawing. May not need to complete the drawing to determine the area by counting (most younger children) or computation (repeated addition or multiplication).	Builds, maintains, and manipulates mental images of composite shapes, structuring them as composites of individual shapes and as a single entity—a row of congruent squares. Applies this composite unit repeatedly and exhaustively to fill the array. If in a measurement context, applies the concept that the length of a line specifies the number of unit lengths that will fit along that line.

Figure 3.1 continues on next page

Developmental Progression	Actions on Objects
Area Conserver. Conserves area and reasons about additive composition of areas (e.g., how regions that look different can have the same area measure) and recognizes need for space-filling in most contexts.	N/A—We removed the area-conserver level from the sequence of the LT for area measurement early in our work on this project, and we therefore did not address the actions on objects to fit it. Chapter 6 addresses this decision.
Array Structurer. With linear measures or other similar indications of the two dimensions, multiplicatively iterates squares in a row or column to determine the area. In multiple contexts, computes area from the length and width of rectangles and explains how that multiplication creates a measure of area. Children also must restructure the region to determine how to use known area measures to find the areas of nonrectangular areas.	Builds, maintains, and manipulates composites (units of units of units) that operate in two dimensions. Mentally composes and decomposes array into rows by columns or individual squares. The mental image may be of a spatial array or at this level especially, a symbolic array.

Figure 3.1. Initial developmental progression of the LT for area measurement (adapted from Sarama & Clements, 2009b).

QUANTITATIVE ANALYSES FOR AREA

As explained in Chapter 1, we employed the Rasch model to verify the developmental progression of the LT for area measurement by using quantitative data from our assessment data. Our findings supported the unidimensionality of the construct of area measurement. The measures, with an eigenvalue of 2.2, explained 61.8% of the raw variance and only 5.3% of the unexplained variance in the first contrast, within acceptable ranges (Linacre, 2005). Evidence of area measurement as a unidimensional construct fulfills one of the underlying assumptions of Rasch modeling, thereby indicating that Rasch modeling is appropriate for this data. The range of MNSQ infit, 0.56 to 1.27, was within the 0.5 to 1.5 range, indicating a satisfactory range for productive measurement (Linacre, 2005); thus, all items on the instrument are well aligned, and we can assume that they are measuring the same construct.

As previously mentioned, item reliability is an estimate of the replicability of item placement within the ordered list of items along the scale and is similar to Cronbach's alpha. Also, item separation indices are estimates of the distance between items on that scale. Item reliability was found to be at 0.98 with a separation index of 7.43. These data suggested that the item-difficulty hierarchy was suitable for the sample used in this study (Bond & Fox, 2007) and that the assessment was able to distinguish between low-performing children and high-performing children. Overall, the item-person map revealed that the distribution of item difficulty corresponded well to the abilities of the children included in our sample. This finding provided further support that the progression of children's abilities in area measurement on the assessment was consistent with the theoretical progression of area development posited in the initial LT for area measurement.

Figure 3.2 depicts all items, with item-difficulty estimates on the *y*-axis and corresponding level intervals marked to represent the range of difficulty for each pair of items designed to assess the same level of understanding following the process used for length. Again, each interval, moving from left to right in the figure, corresponds to an increasingly higher level, as hypothesized in the developmental progression of the initial LT for area measurement.

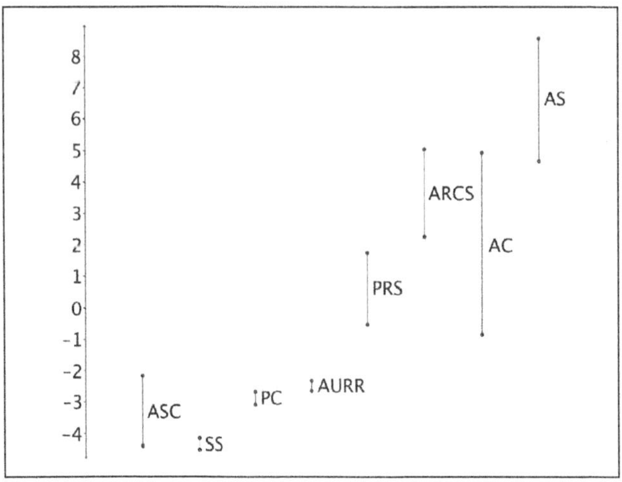

(a)

Figure 3.2 continues on next page

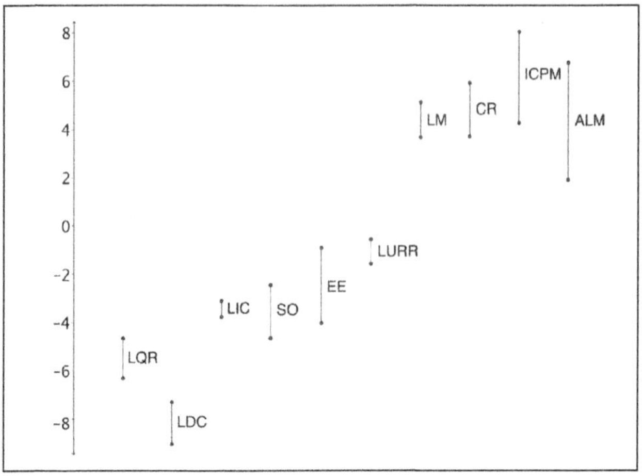

(b)

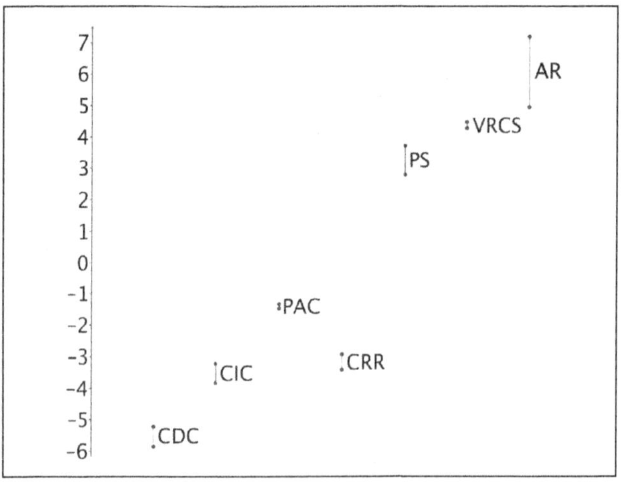

(c)

Figure 3.2. Item difficulty aligned with the levels of the LT for (a) area measurement. ASC = Area Simple Comparer, SS = Side-to-Side Area Measurer, PC = Primitive Coverer, AURR = Area Unit Relater and Repeater, PRS = Partial Row Structurer, ARCS = Area Row and Column Structurer, AC = Area Conserver, and AS = Array Structurer; (b) and (c) show analogous versions for length and volume measurement.

In general, estimates indicate a progression in difficulty of the items that is consistent with our hypothesized developmental progression. That is, an increase in item difficulty corresponds to a move to higher levels in the LT. Two discrepancies are evident, highlighting a need for refinement of the LT for area measurement. The interval for the Area Simple Comparer level begins near the expected point, but the upper range of the interval is well above what we had predicted. Such an overlap may signal the need to eliminate a level that is not distinct from adjacent levels. In addition, the Area Conserver level shows a range that begins below the Partial Row Structurer level and ends below the top of the Area Row and Column Structurer level. Although we designated area comparison and area conservation as specific levels in the initial LT for area measurement, we found that both develop across several levels and in conjunction with other area competencies and concepts, such as spatial structuring.

Chapter 4

Area in the Early Childhood and Primary Grades

Julie Sarama
University of Denver
Douglas H. Clements
University of Denver
Douglas W. Van Dine
Metropolitan State University of Denver
Maria Vukovich
The Center for Victims of Torture
Katherine Newburgh
University of Denver
Jennifer S. McDonel
Radford University
Lisa Napora
Daemen College

Our main objective in this chapter is to chronicle children's growth through the early levels of the developmental progression of the initial learning trajectory (LT) for area measurement. On the basis of our work with children from pre-K through Grade 2, we describe the advent of each level and highlight transitions between these levels, providing examples of children demonstrating thinking and behaviors indicative of specific levels in the LT for area measurement. We also demonstrate that a variety of ways exist to exhibit prominence in one specific LT level. Throughout this chapter, we identify places that suggest a need for refinement, but Chapter 6 discusses in detail the nature of the revisions.

CONSISTENCY WITH INITIAL LT FOR AREA MEASUREMENT: QUALITATIVE EVIDENCE

Pre-Area Quantity Recognizer

At the Pre-Area Quantity Recognizer level, children perceive space and objects within space but demonstrate little specific conceptualization of area or the importance of completely covering a region of space. When researchers ask children at this level to represent the tiling of a rectangular region, the children often draw approximations of circles or other mostly closed figures with no indication of covering the specific region.

On the initial assessment during pre-K, Edith demonstrated development beyond the Pre-Area Quantity Recognizer level; however, Lia and Robert each showed evidence of thinking at the Pre-Area Quantity Recognizer level in several tasks. When the researcher asked Lia to copy a 4 × 5 rectangular array, she drew the boundary line segments and then a cluster of mostly closed shapes inside, as shown in Figure 4.1a; however, she did not show any indication of attempting to cover the region inside the boundary. On the same task, Robert drew six discrete, unconnected, and approximately rectangular shapes in a variety of sizes on the paper, as Figure 4.1b shows. He did not delineate a region with a boundary or attempt to cover the entire region. Although he subsequently demonstrated attributes of more sophisticated LT levels during the initial assessment, his thinking was indicative of the Pre-Area Quantity Recognizer level in this specific task.

a. b.

Figure 4.1. Representations of a 4 × 5 array by (a) Lia and (b) Robert.

Area Simple Comparer

At the Area Simple Comparer level, children may attempt to compare areas by using only one dimension, often by using perceptual objects to compare extent. Our focus children initially used thinking and behaviors at the Area Simple Comparer level after using Pre-Area Quantity Recognizer and before using Side-to-Side Area Measurer. However, to provide a complete picture of this level, we also discuss examples of children using thinking and behaviors at the Area Simple Comparer level as a fallback later in their development, which is consistent with our theoretical perspective of hierarchic interactionalism (see Chapter 1). Further, the evidence of more sophisticated comparisons observed during Grades 1 and 2 confirm the need to revise the level; that is, comparison strategies do not occur at only a single level but develop as a child's understanding of area deepens. Children may base initial comparisons solely on one dimension or overall comparison (e.g., superposition); but as spatial structuring develops, they base area comparisons more on spatial structuring and develop them into comparisons that use linear dimensions.

We saw initial use of thinking at the Area Simple Comparer level in the fall of pre-K. During the initial assessment, the researcher gave Lia a sheet of paper with the outlines of two small 1 × 1 squares and a 3 × 4 rectangle drawn next to one another. The researcher then gave her a square tile, the same size as the small drawn squares and asked, "How many of these squares would fit inside this rectangle?" Although Lia did not demonstrate that she understood how to cover the region, she responded by placing the square tile on top of one of the small squares and saying, "They are the same size." We see this response as evidence of the Area Simple Comparer level because at this level children may compare areas by superimposing one object on the other. In a similar task during a teaching episode in the spring of pre-K, the researcher showed Robert and Edith a picture of a gray square next to a larger rectangle and gave them a square tile that was slightly larger than the square (see Figure 4.2). Instead of superimposing the yellow tile on the gray square, both Robert and Edith used perceptual support and visual comparison to identify the yellow tile as "bigger." When the researcher asked how they knew, both children referred only to the lengths of the vertical sides (one dimension). Again, we see this as evidence of thinking at the Area Simple Comparer level, in which children use only one dimension of figures in their comparisons of area.

Figure 4.2. Edith comparing the areas of the gray square and the yellow tile.

Later in the study, we saw children returning to use thinking at the Area Simple Comparer level in comparing areas even though their understanding of area had progressed much further, which is consistent with our theoretical perspective hierarchic interactionalism. In a teaching episode in the fall of Year 3 (when Lia, who had repeated kindergarten, was in Kindergarten 2 and Robert was in Grade 1), we asked the children to determine whether the top of a student desk or a carpet sample was larger in area. Lia and Robert both demonstrated similar thinking at the Area Simple Comparer level in response to this task. Lia initially held the carpet sample up next to the desktop, aligning the longer side of the rug with the longer side of the desk to compare them. When prompted to cover, she placed the carpet sample on top of the desk and said that the rug was larger in area because it was longer, focusing on only one dimension in her description. Without prompting, Robert placed the carpet sample on top of the desk and aligned the longer and shorter dimensions to compare them, saying, "the rug is bigger because it is hanging off" (see Figure 4.3). Although Robert superimposed the two objects without prompting, he used only one dimension in his comparison.

Figure 4.3. Robert placing the carpet square on the desk to compare areas.

Similarly, in the fall of Grade 2, we presented Robert with the task of arranging six paper rectangles of various sizes and dimensions from smallest to largest by area or by "how much space they cover." He immediately picked out the smallest rectangle. Next, he held up two other rectangles side by side and compared their heights before laying them down next to the smallest one. He proceeded to lay down rectangles one at a time in a line according to their heights. When we asked him why he ordered them as he did, Robert used his fingers to identify the height of the smallest rectangle, describing it as "small." When we asked what he meant, Robert responded, "because it only comes up to here," indicating the height of the square (see Figure 4.4). Further, Robert explained that the rectangles with the same height (although they had different widths) were "the same size." This task was a more complex one than we had previously presented to Robert because we had asked him to order six rectangles according to their areas rather than simply to compare the areas of two rectangles. Again, consistent with our theoretical perspective hierarchic interactionalism, children often demonstrate fallback to using thinking and behaviors from a lower level on more challenging tasks. Robert was using only one dimension to compare the areas of figures (cf. Beilin, 1982), evidence that we believe indicates thinking at the Area Simple Comparer level.

Figure 4.4. Robert arranging rectangles by height when asked to order them by area.

Although Edith showed evidence of more sophisticated levels of thinking in Grade 1 on some comparison tasks, she demonstrated thinking at the Area Simple Comparer level on others. In a teaching episode in the late fall of Grade 1, the researcher gave Edith sheets of paper, each depicting two rectangles, and asked her to compare the areas of each pair (see Figure 4.5). With each, the researcher followed up with a question about which rectangle would contain more square tiles, which were on the table for Edith to see. First, the researcher showed Edith two rectangles, one of which was larger in both height and width (see Figure 4.5a), and asked her to compare their areas. Without using manipulatives, Edith quickly responded, "This one is big" [pointing to the larger rectangle], "and this one is smaller" [pointing to the smaller one].

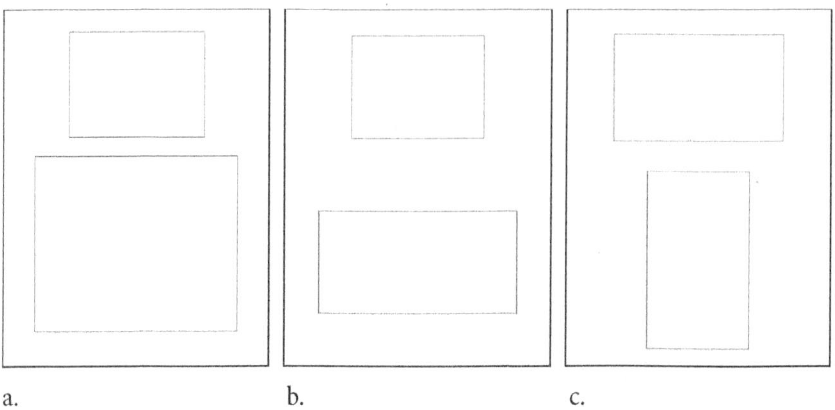

Figure 4.5. Pairs of rectangles used in area comparison task.

When the researcher asked how she knew, she explained, "because it's bigger this way," placing her hands to the left and right of the larger shape. The researcher next presented Edith with two more rectangles; this time the heights were the same, and only the lengths were different (see Figure 4.5b). Again Edith referred only to the length in her comparison, saying that the longer one had a larger area. Edith confirmed in each case that she thought the rectangle she had identified as larger could contain more tiles. As a final task, the researcher asked Edith to compare the areas of two congruent rectangles, one oriented horizontally and the other oriented vertically (see Figure 4.5c). Edith replied, "This one is wide and this one is long," first pointing to the rectangle oriented with the longer dimension horizontally and then to the rectangle oriented with the longer dimension vertically. When the researcher again prompted her to compare the areas of the two rectangles, Edith pointed to the horizontally oriented rectangle stating, "This one is bigger, because it looks wider." When the researcher asked about the number of tiles that each would contain, Edith confirmed that she thought the horizontally oriented rectangle would contain more tiles. Throughout this teaching episode, Edith focused on only one dimension in her comparison of areas, evidence that we saw as indicative of thinking at the Area Simple Comparer level.

Side-to-Side Area Measurer

At the Side-to-Side Area Measurer level, children begin to understand implicitly that regions can cover other regions. When children at this level are drawing, they may not organize, coordinate, or structure two-dimensional space without perceptual support (e.g., tick marks around the perimeter). They may concentrate on only those squares around the inside border of the shape and rely heavily on the sides of the region or on drawing next to existing guides. In attempting to deal with squares in the interior, children have difficulty structuring the space when drawing and counting. Later in this level, children may attempt to fill the region when drawing but leave gaps. They may fail to align drawn shapes. When given enough or more than enough physical tiles, children at this level can complete a covering of a given region, a revision that we discuss further in Chapter 6.

Each of our focus children showed evidence of thinking at the Side-to-Side Area Measurer level in the study. Edith, in pre-K, was the first to demonstrate this thinking; however, Lia (during Kindergarten 1) and Robert (during kindergarten) began showing evidence of this thinking the following year. On the initial assessment, the researcher asked Edith to copy a 4 × 5 rectangular array. She began by drawing the boundary of the rectangle, then used it as a guide to draw approximately square shapes next to one another around a portion of the boundary (see Figure 4.6). Here, Edith focused on only one dimension at a time when she drew across the top and then again when she drew down the right-hand side. Therefore, we did not see evidence of her coordinating in two dimensions. She was not attending to the interior space, nor did she concern herself with making the drawn shapes equal in size. We saw this as evidence of thinking at the Side-to-Side Area Measurer level.

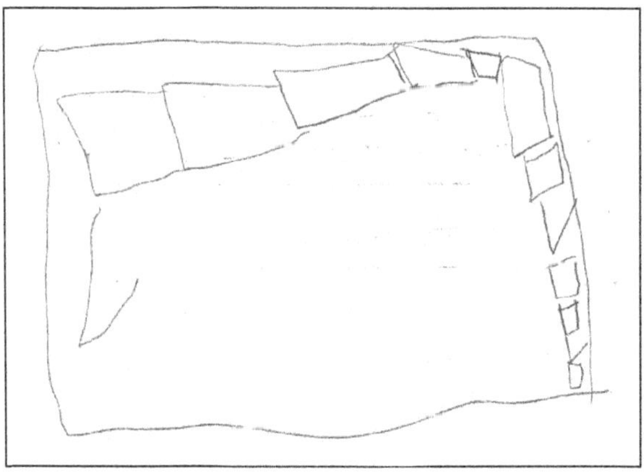

Figure 4.6. Edith's representation of a 4 × 5 rectangular array.

One task that we found to be useful in eliciting behaviors that allowed us to better interpret the children's mental actions on objects was providing children with a partially filled array and asking them to draw to complete the array. Because that task was useful in assessing multiple levels of the LT, we posed versions of the task during multiple teaching episodes. We first presented Robert a task like that one in the spring of his kindergarten year. In his attempt to complete a partially erased 5 × 6 array, he left gaps and misaligned his drawn shapes (see Figure 4.7). Robert did not apply an explicit understanding that he needed to cover the entire region or align drawn squares in rows, as evidenced by his leaving gaps between his drawn shapes. Therefore, we saw his thinking as indicative of the Side-to-Side Area Measurer level.

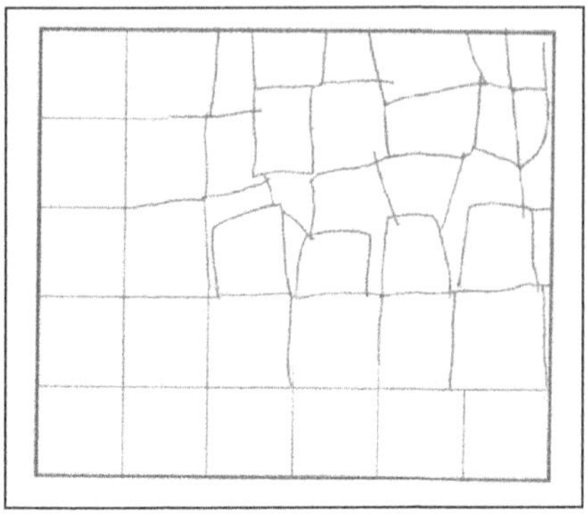

Figure 4.7. Robert's attempt to complete a partially erased 5 × 6 array.

Lia first demonstrated thinking at the Side-to-Side Area Measurer level in similar drawing tasks during the spring of Kindergarten 1. She routinely left gaps or used improper alignment of drawn shapes, consistently choosing to represent only certain aspects of the structure. We gave Lia the same task that we have described for Robert—completing a partially erased 5 × 6 array. Initially, she attended to the partially drawn squares in the second column and drew mostly closed individual shapes to complete any that were not entirely represented; however, she did not pay attention to size or coordination of dimensions. Although her attempt to complete the remainder of the rectangle indicated that she had a partial understanding that she needed to fill the space, she left gaps between her drawn shapes and did not completely cover the area. In addition, she did not coordinate or align the drawn shapes in rows (see Figure 4.8). We saw this response as indicative of thinking at the Side-to-Side Area Measurer level.

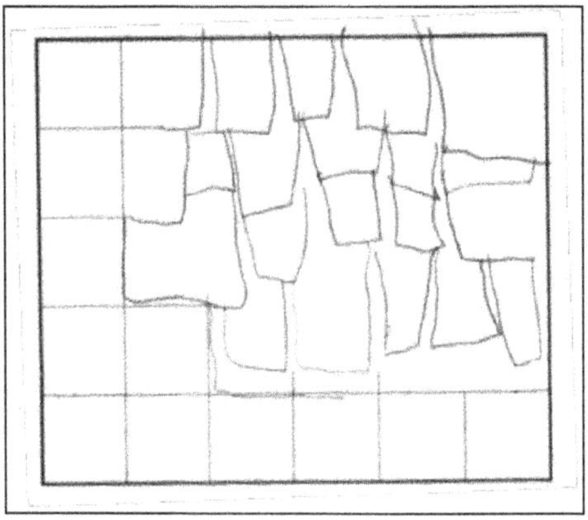

Figure 4.8. Lia demonstrated thinking at the Side-to-Side Area Measurer level in her drawing to complete a partially erased array.

During the next teaching episode, the researcher presented Lia with the same partially erased 5 × 6 array, but instead of drawing, the researcher asked her to estimate the number of tiles needed to completely cover the rectangle. Lia counted only the completely drawn squares, counting down the left column then across the bottom, to arrive at a total of 11. The researcher asked if those 11 squares would cover the whole space, and Lia confidently responded, "Yes." When the researcher asked specifically whether that included the open space, Lia responded with, "Watch" and then started counting again, this time counting down the left column and then moving to a pattern with little observable structure that included the bottom squares and the open area. Her total on this counting was 16. When the researcher followed up by asking her to show how she counted, Lia began again with the first column and then counted in a slightly different pattern, arriving at a total of 17. She showed no concern that she had obtained three different counts and in fact, showed confidence that her counting had been correct each time. We saw this response as evidence of thinking at the Side-to-Side Area Measurer level because her focus was on counting the squares already drawn; and when pressed to focus on the portion not covered, she showed little indication of structuring the space.

Lia continued to exhibit thinking at the Side-to-Side Area Measurer level in her drawings through the following school year. During a teaching episode in the third year of the study (Lia's Kindergarten 2), the researcher gave her a 1 × 1 square tile and asked how many she would need to cover a 3 × 4 rectangle. She began by placing the tile in the upper-left corner of the rectangle, then pointed and counted around the inside of the boundary of the rectangle, arriving at a total of 15. She only counted around the border and did not attend to the interior space in her counting—a behavior consistent with the Side-to-Side Area Measurer level. Next, the researcher asked Lia to tile the rectangle with 1 × 1 tiles, and she was able to accomplish this task correctly. The reviewer subsequently asked her to draw a picture of her completely tiled rectangle. Her drawing consisted of closed shapes, some rectangular and others circular, in a single row across the rectangle, with gaps between the shapes (see Figure 4.9a).

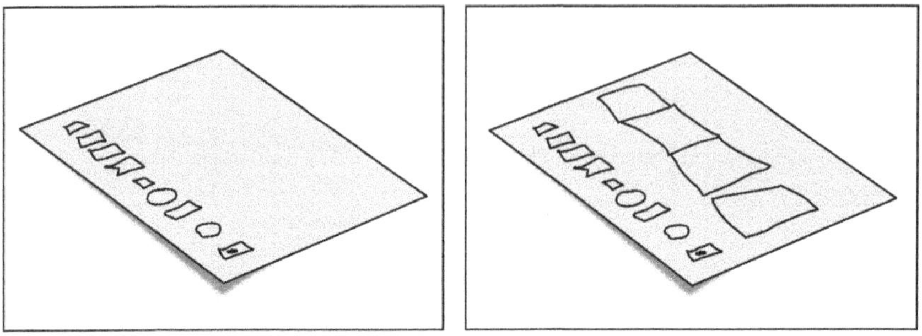

a. b.

Figure 4.9. Lia's first (a) and second (b) attempts to represent a tiled rectangle.

The researcher followed up by prompting Lia to try to draw squares the same size as the tiles that she had used. This attempt resulted in her drawing a row of four somewhat rectangular shapes that were approximately the same size as the tiles. However, she still left a gap between her first and second drawn shapes, and her drawing product did not indicate that she had attempted to completely cover the space (see Figure 4.9b). With guidance, Lia appeared to make a drawing that better represented the tiles that she had used to cover the rectangle. However, she did not demonstrate an understanding that the drawing needed to completely cover the rectangle or that the size and shape of drawn tiles should be uniform. Instead, she drew discrete shapes in a single line. Lia's reliance on existing guides in her counting, as well as the drawings in which she attempted to represent a completely tiled rectangle, provided evidence of thinking at the Side-to-Side Area Measurer level.

While the focus children progressed through the Side-to-Side Area Measurer level, they began to recognize when they had fully covered a specific region; however, their drawing and counting suggests that they relied heavily on existing guides. They routinely became disorganized when moving into the interior and away from the boundary. This outcome was evident in routine counting errors, as well as in their focus on only one dimension of a specific region.

Primitive Coverer

Development in spatial structuring characterizes movement into the Primitive Coverer level. Children at this level are able to recognize a complete covering of a region and have developed an implicit visual patterning that allows them to guide and connect the placement of squares in approximate rows. In counting tasks, children at the Primitive Coverer level fully attend to the interior space but often count first around the boundary and then count unsystematically in the interior, possibly skipping some squares and double-counting others. In drawing tasks, children have progressed past drawing mostly closed individual squares to using existing sides to guide placement, thereby creating concatenations of squares that are approximately aligned in rows. Although children have become more cognizant that a correct tiling should include no gaps, they are not fully aware of the need for equal-sized units while tiling or measuring a specific region to find the area.

As early as the initial assessment, we saw evidence that Edith was beginning to recognize the interior space, but it was not until kindergarten that she showed definitive evidence of thinking at the Primitive Coverer level in both counting and drawing. During a teaching episode in kindergarten, when the researcher asked her to determine the total number of 1×1 tiles needed to cover a 5×6 rectangle in which the first column and bottom row had already been drawn, Edith counted around the border then unsystematically in the interior, as shown in Figure 4.10a, to arrive at a total of 25.

After Edith had counted, the researcher asked her to "fill in the rest of the missing squares" by drawing. She drew closed units, approximately aligned with those initially drawn in the rectangle. However, when she reached the upper interior space (farther away from the existing guides), Edith began to make alignment errors, as shown in Figure 4.10b. On this task, we saw her attempt to completely cover the space as a movement beyond the Side-to-Side Area Measurer level and indicative of thinking at the Primitive Coverer level.

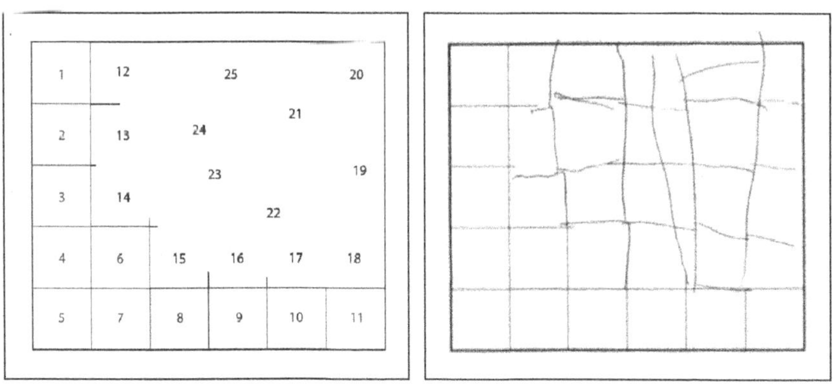

a. b.

Figure 4.10. Given a partially erased array, Edith demonstrated thinking at the Primitive Coverer level while counting (a) and drawing (b).

As another example, in spring of Grade 1, the researchers provided Robert with a selection of tiles (specifically, squares in two different sizes and rectangles in two different sizes) and asked him to tile a 3×4 rectangle by using the size of the smaller square. The researchers gave Robert more than enough of each size and shape of tile to cover the whole space. He tiled in rows, using 12 equal-sized square tiles, forming a 3×4 array that fit exactly within the rectangle. However, even with prompting from the researcher, Robert was unable to articulate why all tiles needed to be the same size. We identify his thinking to be consistent with the Primitive Coverer level rather than a more advanced level, because he did not yet demonstrate comprehension of the importance of unit size in measuring area.

Lia showed evidence of thinking at the Primitive Coverer level in the spring of Kindergarten 2. During a teaching episode at that time, the researcher gave her a selection of different objects that she could use as tiles and asked her to find the area of a small rug. She first selected two plastic squares, placing them side by side in one corner of the rug. Next, she selected some larger manila envelopes and began placing them on the rug, adjusting the orientation of each envelope as she placed it so that she would completely cover the rug. After she was done covering, Lia unsystematically counted what she had placed, counting envelopes and squares equally and double-counting at times. Similarly, in the next teaching episode, when the researcher asked Lia to find the area of the bottom of a box, she articulated, "I want the whole bottom to be covered" while she tiled. In both of these episodes, Lia demonstrated an explicit understanding that she needed to cover the entire region, but she lacked an understanding of unit, which is consistent with the thinking and behaviors at the Primitive Coverer level.

Later in the spring of Kindergarten 2, we saw another example of Lia exhibiting behaviors indicative of the Primitive Coverer level. The researcher gave her a partially erased 5×7 array (see Figure 4.11a) and asked Lia to draw what it would look like if it were completely covered. She started by drawing two continuous horizontal line segments to complete the first two rows (see Figure 4.11b), then delineated the squares in those rows by drawing vertical line segments that anchored each originally drawn vertical line to the second horizontal line that she had drawn. She subsequently drew a horizontal line segment to delineate the next row, then went back to drawn vertical line segments within the row, anchoring each to the row above it before moving on to drawing the next horizontal line (cf. Miller, 2013). In her first completely drawn row, she added an extra vertical line segment that created an additional column, which she proceeded to include in each subsequent row. Again, we saw this outcome as evidence that she was trying to fill the space without focusing on equal-sized units. Because she applied an explicit understanding that she needed to cover the entire region but did not maintain equal-sized units while drawing, we interpreted her drawing as indicative of thinking at the Primitive Coverer level.

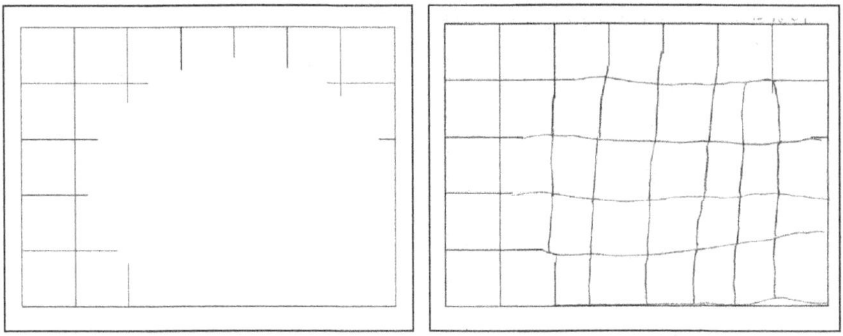

a. b.

Figure 4.11. Given a partially erased array (a), Lia demonstrated thinking at the Primitive Coverer level (b).

Area Unit Relater and Repeater

At the Area Unit Relater and Repeater level, children draw without gaps or overlaps. They appreciate the need for identical units both in drawing and in tiling. Children are able to measure by repeated use of a unit (called *iteration*), although they may not always be precise in their iterations. They are also able to relate the size and number of units explicitly and understand that they need a greater number of smaller units if they want to cover a given area. A sense of two dimensions, recognizing but not necessarily coordinating both dimensions, guides children's thinking about area at the Area Unit Relater and Repeater level. A distinction between Area Unit Relater and Repeater and Primitive Coverer (as well as Side-to-Side Area Measurer) is that instead of drawing and counting unsystematically in the space, the children use rows or columns as an intuitive structure to guide their drawing and counting because they recognize the need for equal-sized units. However, they have not yet begun to see a row or column as a unit of units; and therefore, when a researcher asks them to find the area of a figure, they rely on counting individual units. On the basis of our observations and findings, the need for further refinement of the Area Unit Relater and Repeater level became evident. We emphasize several refinements in the following examples; however, we specify the nature of these refinements in Chapter 6.

Robert first displayed evidence of thinking at the Area Unit Relater and Repeater level in the spring of kindergarten. When the researcher asked him to cover a desk and gave him only two square tiles, Robert placed them adjacent to one another in the top-left corner of the desk. When the researcher asked him how many more he would need, Robert iterated with his finger along the horizontal edge of the desk and correctly determined that he would need four more square tiles to complete the row. When the researcher asked how many rows he would have, he stated that he would have three, indicating that he would need that number of tiles along the vertical edge of the desk. Although he gave no explicit evidence of seeing the row as a unit of units, Robert traced with his finger across each horizontal row, demonstrating that he was attempting to completely cover the space. Consistent with thinking at the Area Unit Relater and Repeater level, he maintained the image of each placement while he counted individual iterations along the first row and column.

During the fall of Grade 1, Robert again showed evidence of thinking at the Area Unit Relater and Repeater level when the researcher gave him a partially erased 5 × 7 array and asked him to draw what it would look like if he completely covered it. Initially, with some prompting, he completed the first row by drawing the missing sides of each square unit. The researcher then prompted him to think about area in terms of rows and columns instead of individual squares. Aided by this guidance, Robert used a ruler as a straightedge to draw vertical line segments to match the size of each square unit in the top row; however, he then returned to drawing small horizontal line segments to demarcate individual squares and complete each column (see Figure 4.12). We saw this outcome as evidence that he was not yet recognizing the columns as units of units, again demonstrating thinking and behaviors consistent with the Area Unit Relater and Repeater level because he drew approximately equal-sized units one at a time.

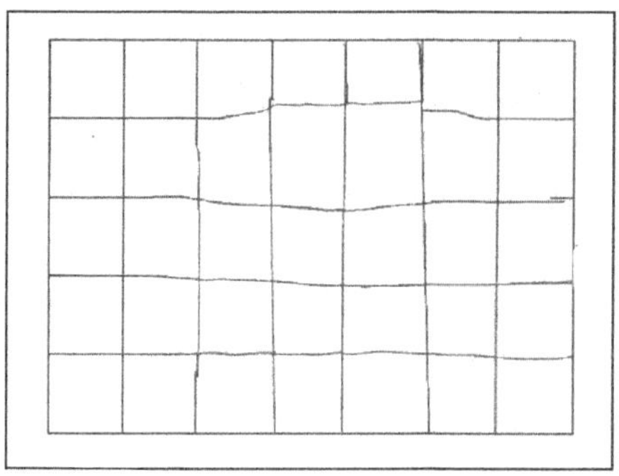

Figure 4.12. Robert drew small horizontal line segments to draw individual squares to complete a partially erased array.

It was not until Grade 1 that we characterized Lia as demonstrating thinking that was dominantly at the Area Unit Relater and Repeater–level. In a teaching episode in the fall, the researcher presented Lia with a 4 × 5 rectangle with only the first column of squares partially drawn in. When the researcher gave her a pencil and asked her to represent what the rectangle would look like completely covered with squares the size of those already drawn, Lia began by drawing units to complete the first horizontal row. The researcher stopped her at that point and asked, "What if we just had that much; can we figure out how much it would be if the rest was filled in?" Lia pointed and counted in the empty space in the second and third vertical columns, indicating four squares in each column. When the researcher asked for a total, Lia pointed and counted individual squares, including those already drawn, and used columns as an intuitive structure while she iterated in the empty space. Although Lia miscounted in the fourth and fifth columns by one tile each and arrived at a total of 22 because she counted individual tiles, using columns as an intuitive structure, we interpreted this result as evidence of thinking at the Area Unit Relater and Repeater level.

We saw Edith's initial foray into the Area Unit Relater and Repeater level in the spring of kindergarten, when the researcher gave her two square tiles and asked her how many she would need to cover a small rug. She began by placing the two square tiles side by side in the top left-hand corner. She then iterated the two tiles, using a "leapfrog" strategy along the top edge of the rug, counting while she iterated each tile. On reaching the end of the row, she moved both tiles back to the left edge of the rug where the next row would begin and used the same leapfrog strategy to iterate along that row. She continued this process until she had covered the entire area of the rug. The ability to iterate a unit to cover an area is evidence of thinking at the Area Unit Relater and Repeater level. We saw with both Robert and Edith that this task, involving two tiles and a specified region, elicited thinking at the Area Unit Relater and Repeater level in children; we therefore believe that this task is a useful one for assessing thinking at this level.

In the fall of Grade 1, Edith again provided evidence of the repeating (or iterating) aspect of the Area Unit Relater and Repeater level. The researcher gave her a 4 × 6 rectangle on which the top row and left-hand column of squares were already drawn and asked Edith, "How many tiles do you need to fill in this rectangle?" When the researcher gave her a physical tile, Edith iterated over the squares already drawn, counting to 9. Unprompted, she then iterated through the empty space by using an intuitive row structure, counting each unit while she iterated to determine that 15 tiles would cover that area. As with Robert and Lia, Edith's ability to rely on the rows as an implicit marker to keep track of her iterations demonstrated that she was exhibiting thinking and behaviors indicative of the Area Unit Relater and Repeater level.

Edith again showed evidence of thinking at the Area Unit Relater and Repeater level in Grade 2. In a teaching episode in the fall of Grade 2, the researcher gave Edith a 4 × 5 rectangle with tick marks along the sides to indicate the dimensions and asked her to find the area. She quickly articulated, "You take one, put it in this square; and as you are doing it, count them, and keep doing it across like this" while she iterated a single tile between each pair of tick marks to correctly reach a total of 5 in the first row. Next, she tiled the entire first row and then used the row structure as a guide to mentally iterate individual units through the rest of the region, thus determining a total area of 20. We found tasks like this one that included tick marks were useful in encouraging children to maintain equal units in both tiling and drawing.

Partial Row Structurer

At the Partial Row Structurer level, children can build, maintain, and manipulate mental images of units of units. They can structure two-dimensional space by using single units, as well as by using a unit of units. Children at this level recognize a row (or column) as a unit of units and can apply that unit of units repeatedly, but not necessarily exhaustively, because intuition guides its application. Children illustrating thinking at the Partial Row Structurer level often use additive reasoning or skip counting to determine the area of a region. Lia (in the third year of the study) and Edith (in the final year of the study) both showed evidence of thinking at the Partial Row Structurer level. In contrast, Robert did not appear to recognize a unit of units or apply additive reasoning to reach a total during the study. He did, however, show preliminary evidence of thinking at the Partial Row Structurer level in the final year of the study.

During the fall of Grade 2, the researcher asked Robert to determine the area of a 4×5 region in which part of the left column and the top row had already been drawn. When the researcher asked him how many tiles would cover the entire area, Robert tentatively conjectured, "Maybe there is five in each row?" When the researcher gave him several physical tiles but not enough to cover the region, he laid out four tiles to complete the second row. He then counted one through five in each row, counting both the tiles already drawn and iterating in the empty space. We saw that Robert could visualize rows but could not operate on a row as a unit of units; he therefore only counted individual units during the task. This outcome demonstrated that although he was reaching into the Partial Row Structurer level, he was still dominantly using thinking at the Area Unit Relater and Repeater level.

During Grade 1, Edith provided a clear example of thinking at the Partial Row Structurer level when the researcher gave her a partially erased 5×7 array and asked her to draw a complete covering. Edith drew a vertical line segment to identify the second column and then drew individual horizontal line segments to identify each square in that column. We noticed that Edith's thinking and behaviors were similar to those of Robert on this task, and we took this outcome as evidence that she was thinking at least at the Area Unit Relater and Repeater level. Edith next drew vertical line segments from the top to the bottom of the rectangle to mark each of the remaining columns before turning her paper 90 degrees and drawing line segments across the rectangle to mark off each row (see Figure 4.13). When the researcher asked for a total, Edith recognized and articulated that each column included five squares and attempted to count by fives (although demonstrating counting difficulty at the decade mark): "5, 10, 20, 30, 40, 50, 60," pointing at each column from left to right while she counted. With guidance, she was able to correct her counting and obtain a correct area of 35, using a similar skip-counting method. Although she struggled with counting by fives, Edith demonstrated a conceptualization of a column as a unit of units and was able to apply additive reasoning to obtain a total area. We see this result as evidence of thinking at the Partial Row Structurer level.

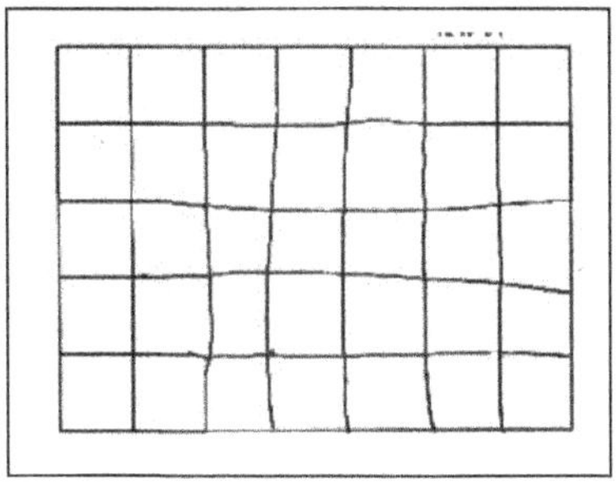

Figure 4.13. Edith drew parallel line segments to complete a partially erased array.

Similarly, during the fall of Grade 2, Edith further demonstrated thinking at the Partial Row Structurer level. When the researcher gave her a 4 × 6 rectangular piece of paper, called it a *candy bar,* and asked her to indicate what half of it would be, Edith used her finger to trace vertically down the center of the candy bar and replied, "I just have to see how many would go in a row." Next, she was able to see that four squares would be in each row and then used additive reasoning, counting in groups of four (4, 8, 12), to determine the number of squares in half the candy bar. She stated, indicating each half of the paper, "Then you split it right here" [gesturing with her finger vertically down the center again] "so I still get three rows and you would get three rows, too." The researcher probed, "three rows of how many?" and Edith immediately said "four." On this task, Edith demonstrated the ability to independently visualize square units and use rows as a unit of units in her conceptualization of area, exhibiting thinking consistent with the Partial Row Structurer level.

The researcher gave Lia a blank 4 × 5 rectangle in a teaching episode during spring of her Grade 1 year; and she was able to measure with a ruler to determine the dimensions and relate those dimensions to the number of units along each side, saying, "five blocks this way" [aligning the ruler across the top of the rectangle], "four blocks this way" [aligning the ruler along the right side], "five blocks" [aligning the ruler along the bottom], "and four blocks" [aligning the ruler along the left side]. Next, she laid out five 1 × 1 tiles to complete the top row and immediately said that there would also be five tiles along the bottom. The researcher asked, "If there were five along the top and five along the bottom, how many would that be?" to which Lia responded, "10 tiles." The researcher then asked if that amount would that fill up the whole shape, and Lia answered, "No, we would need five here" [tracing her finger across where the second row would be] "and five more" [tracing across where the third row would be]. Although Lia built and manipulated a unit of units, in this case a row, we do not have evidence of her using the unit of units exhaustively. This evidence was consistent with thinking at the Partial Row Structurer level.

Levels More Sophisticated Than Partial Row Structurer

At the Northeast site, we saw some children beginning to reach into the Area Row and Column Structurer level. However, these children did not consistently demonstrate thinking at levels more sophisticated than Partial Row Structurer. Chapter 5 includes discussion of the more sophisticated levels of the LT for area measurement and the work with older children at the Midwest site.

LONGITUDINAL GROWTH CHARTS

To help summarize the preceding qualitative data, we present individual longitudinal growth charts for the three focus children that this chapter describes—Robert, Edith, and Lia—for each of whom we have 4 years of data. In each chart, vertical line segments represent the levels of thinking demonstrated during each teaching episode involving area measurement. In addition, a shaded rectangle represents what we identify as the dominant level of thinking that the child demonstrated during that teaching episode (see Chapter 2, page 45, for a more detailed description of the growth-chart components). Because the qualitative descriptions above have already presented details, our purpose here is to furnish growth charts with brief summaries and interpretations of what each growth chart presents.

Edith's Growth

At the beginning of the study, we categorized Edith's dominant level of thinking in pre-K as Side-to-Side Area Measurer (see Figure 4.14). In kindergarten, she showed some progression in spatial structuring ability, resulting in her dominant level of thinking moving to the Primitive Coverer level. Throughout Grade 1 and Grade 2, Edith demonstrated recognition of a unit as well as the ability to build and act on a mental unit of units. In other words, she demonstrated a conceptualization of a row as a unit of units and was able to apply additive reasoning to obtain a total area. During those 2 years, her dominant levels of thinking were Area Unit Relater and Repeater and Partial Row Structurer, not consistently one or the other but shifting between the two depending on the complexity of the task (with greater complexity often leading her to use thinking and behaviors at the Area Unit Relater and Repeater level). This shift between two levels is consistent with our theoretical perspective of hierarchic interactionalism. During one teaching episode in the fall of Grade 2, Edith's dominant level of thinking was Primitive Coverer, which seems out of place. We attributed this apparent fallback to the limitations of the tasks presented rather than a fallback in thinking. That is, area-measurement tasks presented during that teaching episode did not prompt children to demonstrate higher levels of thinking; children completing those tasks completely and correctly demonstrated thinking at the Primitive Coverer level.

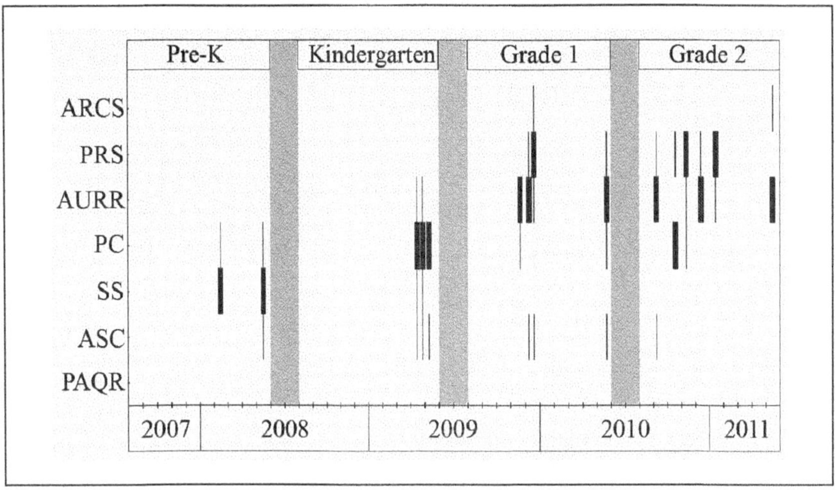

Figure 4.14. Edith's longitudinal growth chart. PAQR = Pre-Area Quantity Recognizer, ASC = Area Simple Comparer, SS = Side-to-Side Area Measurer, PC = Primitive Coverer, AURR = Area Unit Relater and Repeater, PRS = Partial Row Structurer, and ARCS = Area Row and Column Structurer.

Robert's Growth

On the initial assessment during pre-K, Robert's dominant level of thinking appeared to be Area Simple Comparer, but his thinking appeared to shift toward the Side-to-Side Area Measurer level while he was still in pre-K (see Figure 4.15). In kindergarten, Robert continued to develop a more sophisticated understanding of area measurement, and his dominant level of thinking was Primitive Coverer or Area Unit Relater and Repeater, depending on the task. Like Edith, and consistent with hierarchic interactionalism, more difficult tasks resulted in Robert's use of thinking and behaviors at a less sophisticated level (i.e., at the Primitive Coverer level) even though he had shown competence at the Area Unit Relater and Repeater level. By Grade 1, he had developed a strong sense of unit and the ability to maintain unit size in area tasks and was predominantly at the Area Unit Relater and Repeater level. During Grade 2, thinking at the Area Unit Relater and Repeater level continued to be dominant, but he did show one foray into Partial Row Structurer early in the school year. Again, we want to point out the apparent fallback in the fall of Grade 2. As we explained in our discussion of Edith's chart, we attributed this outcome to the limitations of the tasks presented rather than a fallback in Robert's thinking about area measure.

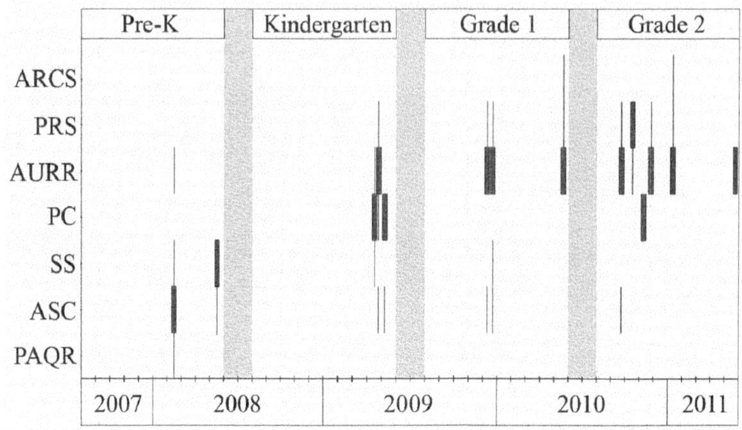

Figure 4.15. Robert's longitudinal growth chart. PAQR = Pre-Area Quantity Recognizer, ASC = Area Simple Comparer, SS = Side-to-Side Area Measurer, PC = Primitive Coverer, AURR = Area Unit Relater and Repeater, PRS = Partial Row Structurer, and ARCS = Area Row and Column Structurer.

Lia's Growth

Like Robert, Lia demonstrated Area Simple Comparer as her dominant level of thinking on the initial assessment; but later that year, her dominant level of thinking had progressed to the Primitive Coverer level (see Figure 4.16). Although Lia never demonstrated Side-to-Side Area Measurer as her dominant level of thinking during a teaching episode, she did demonstrate thinking at the Side-to-Side Area Measurer level during both teaching episodes in pre-K. Because of the nature of the teaching-experiment methodology and the frequency of our sessions with the participants in the study, we may have missed the point at which her thinking was dominantly at the Side-to-Side Area Measurer level because she quickly moved into the Primitive Coverer level. We categorized Lia's dominant level of thinking as Primitive Coverer during the second and third years of the study, during her 2 years in kindergarten. When she entered Grade 1 in the fourth year of the study, her understanding of an area unit became more solidified, as indicated by her dominant level of thinking, for the most part moving to the Area Unit Relater and Repeater level, with one foray into the Partial Row Structurer level toward the middle of the year.

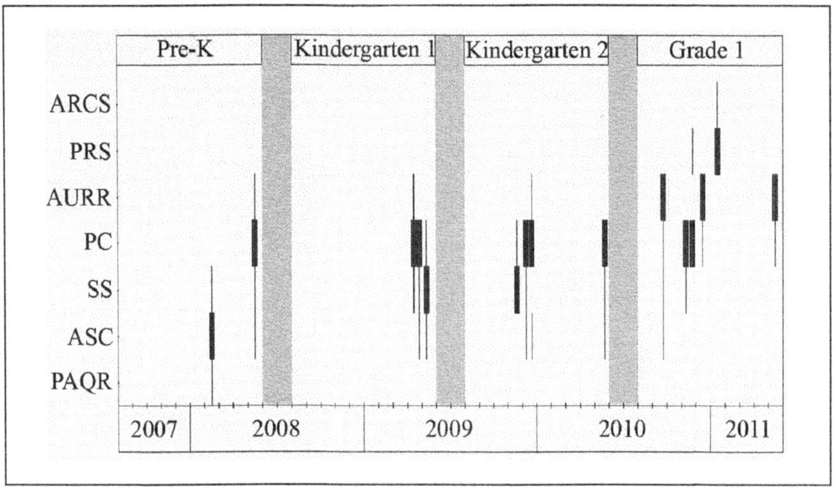

Figure 4.16. Lia's longitudinal growth chart. PAQR = Pre-Area Quantity Recognizer, ASC = Area Simple Comparer, SS = Side-to-Side Area Measurer, PC = Primitive Coverer, AURR = Area Unit Relater and Repeater, PRS = Partial Row Structurer, and ARCS = Area Row and Column Structurer.

DISCUSSION OF LONGITUDINAL GROWTH CHARTS

In Figure 4.17, we overlay the charts for Edith, Robert, and Lia to present a composite longitudinal growth chart. From this chart, we can see a general positive growth trend while the three focus children progressed in their measurement thinking from pre-K through Grade 2. Thus, our findings provided support of the developmental progression in the initial LT for area measurement.

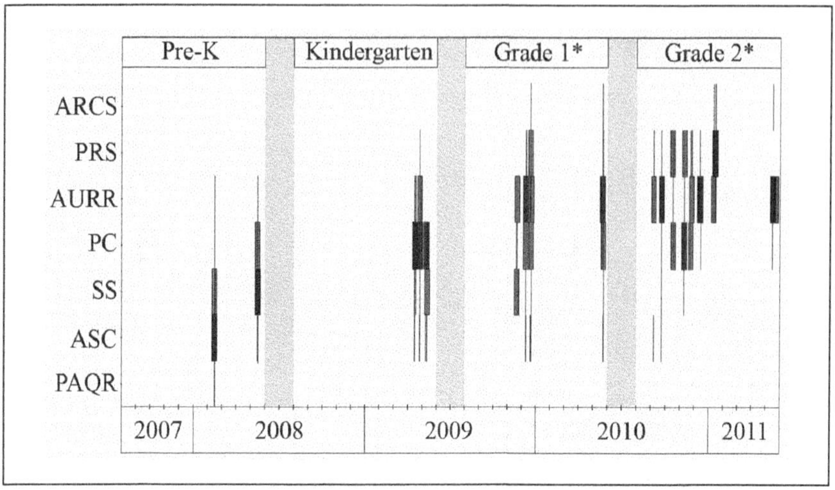

Figure 4.17. Composite longitudinal growth chart. Asterisks on Grade 1 and Grade 2 headings indicate that Lia's data from her second year of kindergarten appears in the Grade 1 column, the third year of our study; and Lia's Grade 1 data appears in the Grade 2 column. PAQR = Pre-Area Quantity Recognizer, ASC = Area Simple Comparer, SS = Side-to-Side Area Measurer, PC = Primitive Coverer, AURR = Area Unit Relater and Repeater, PRS = Partial Row Structurer, and ARCS = Area Row and Column Structurer.

By investigating the growth of our focus children over the course of 4 years, we found overall growth patterns that validated components of the initial LT for area measurement. Overall, the children progressed from thinking at the Area Simple Comparer level to thinking at the Partial Row Structurer level, spending varying amounts of time functioning simultaneously at two contiguous levels. Edith and Robert exhibited consistent but not constant growth throughout the longitudinal study. However, Lia displayed greater variability. As Lia's longitudinal growth chart indicates, her thinking appeared to plateau during Kindergarten 1 and Kindergarten 2. This finding is cause for further investigation. Although the children showed general growth trends that validated the initial LT for area measurement, we found evidence that suggested a need for revision. We discuss our revised LT for area measurement in Chapter 6.

Chapter 5

Area in the Middle and Later Elementary Grades

Jeffrey E. Barrett
Illinois State University
Craig J. Cullen
Illinois State University
Amanda L. Miller
Illinois State University
Cheryl L. Eames
Southern Illinois University, Edwardsville
Melike Kara
Towson University
David Klanderman
Trinity Christian College

In this chapter, we have two main goals. First, we describe the levels of the developmental progression of the initial learning trajectory (LT) for area measurement (Sarama & Clements, 2009b) exhibited by the children in the Midwest cohort while they progressed from the Primitive Coverer level and into the Array Structurer level. We offer exemplars of children's work and discuss their conceptual understanding of area units, including operations with area units. Second, we describe children's learning within and across specific levels of the LT, including their developmental transitions between successive levels of thinking. To address this goal, we describe the progress of one child, Anselm, through each of the levels. His path of development illustrates an empirical storyline that is based on the longitudinal data showing a plausible sequence of levels. We also provide auxiliary examples from four additional children who exhibited behaviors and thinking that were indicative of specific levels. Throughout this chapter, we identify places that indicate a need for modification to the overall LT. Finally, we present a summative collection of growth charts describing the progress of five children throughout the longitudinal study (3 to 4 years, depending on attrition) at the Midwest site.

CONSISTENCY WITH THE INITIAL LT FOR AREA MEASUREMENT: QUALITATIVE EVIDENCE

Primitive Coverer

As explained in Chapter 4, children at the Primitive Coverer level draw complete coverings but have some difficulty aligning objects while they work to cover a region. In that chapter, we presented evidence that children began making primitive coverings as early as kindergarten and continued the process into Grade 2. Because we started working with the children in the Midwest cohort in Grade 2, most of our focus children had already progressed beyond the Primitive Coverer level. For the children in the Midwest cohort who had not yet progressed beyond that level, the manifestation of their thinking at the Primitive Coverer level differed slightly from the Primitive Coverer level thinking for the children in the Northeast cohort. Subsequent discussion describes persistent use of thinking at the Primitive Coverer level beyond Grade 2 for one focus child, Drew. But we first discuss Anselm, who exhibited the Primitive Coverer level of thinking during the initial assessment in Grade 2 before illustrating more sophisticated levels.

Anselm was one of the few focus children who had not yet progressed completely beyond the Primitive Coverer level by Grade 2. On the initial assessment in the spring of Grade 2, the researcher asked Anselm to determine the number of square units in a partially erased 5 × 6 rectangular array. He attended to the existing squares by counting around the border, then worked unsystematically in the interior in an approximately L-shaped pattern, pointing and counting aloud from 1 to 46 (see Figure 5.1). We interpreted this strategy as indicative of the Primitive Coverer level because he was very unsystematic in his counting of individual shapes, yet he demonstrated an understanding that he needed to cover the entire region. Interestingly, this initial assessment was the only time that Anselm exhibited thinking at the Primitive Coverer level during the 4-year longitudinal study.

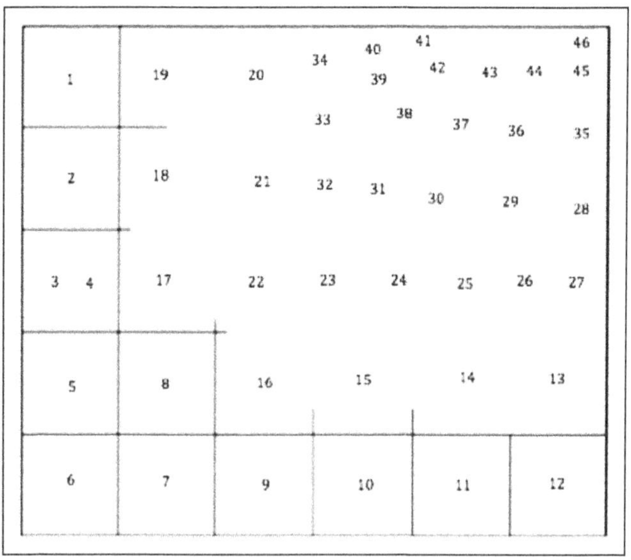

Figure 5.1. Anselm exhibited a counting strategy that was consistent with behaviors and thinking at the Primitive Coverer level in a partially erased array.

Drew was another child in the Midwest cohort who had not yet progressed beyond the Primitive Coverer level by Grade 2. However, Drew's thinking and behaviors that were indicative of the Primitive Coverer level were often novel. He carefully attended to the interior space of a rectangular region and demonstrated an explicit understanding that he needed to cover the entire region. To do so, Drew invented a rule for quantifying the area by counting around a border ring and then counting around further interior rings—each successive rectangular ring moving inward contained half the number of units that the previous ring included. He reported the sum of the rings as the area of the rectangular region. He applied this numeric approach to produce a measure of area several times during Grades 4 and 5.

For example, when asked to determine the area of a 6 × 6 square, as shown in Figure 5.2a, Drew discussed "painting" the outside first. He traced around the square region just inside the boundary and explained, "I was trying to do it like painting around the outside first . . . so that would be like 24 centimeters already." Our interpretation is that he counted the length of each row of squares with an exterior edge, finding six edge lengths of tiles along each edge of the rectangle, for a total of 24 "centimeters." He continued tracing as he explained that "half of 24 is 12 . . . so 24 plus 12 is 36 I think . . . and then half of 12 would be six . . . so 42 . . . and then half of six would be three . . . so that would be 45." In other words, he computed half of 24 because he expected to find 12 units along the ring just inside the first border ring (which he had already asserted to have 24 squares). Next, he computed half of 12 to quantify the next interior ring and then computed half of 6 to quantify the last interior ring. He added 24, 12, 6, and 3 to report an area of 45 (see Figure 5.2b for our depiction of Drew's use of his border-rings "rule"). After reporting an area of 45, Drew said, "and maybe it would cover it all up by then." His use of the word *maybe* to qualify his claim indicated a lack of certainty.

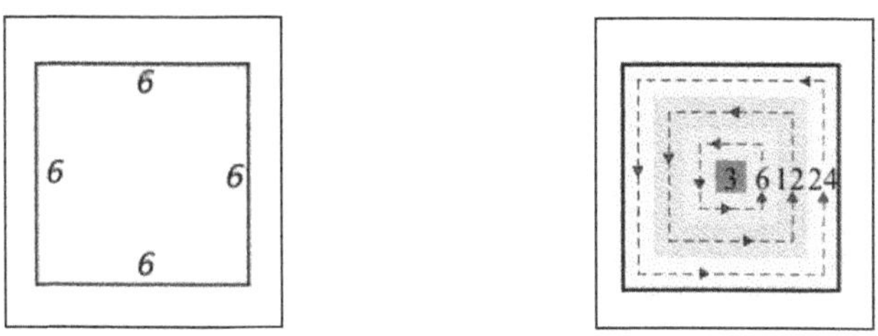

a. b.

Figure 5.2. The 6 × 6 square task (a) on which Drew demonstrated thinking at the Primitive Coverer level in his explanation of "painting around the outside" (b).

Drew used a purely numeric approach on this task, but he did not show the location of the units within each of the rings in the square region, nor did he explain how those numbers corresponded to locations in the region. Unfortunately, we did not ask him to draw the 24 "centimeters" with which he "painted" the first ring, nor did we ask him to show us any of the 45 he reported as the area. Hence, we cannot say whether Drew was counting squares or some other shape of units; we can merely report that he was summing a set of rings that he expected to cover the region. Interestingly, he used a form of this border-ring approach sporadically during Grades 2, 3, and 4, as well as briefly at the beginning of Grade 5, even in cases where it was not adequate for a given task.

Although children at the Primitive Coverer level commonly count around the border of a region, these children then count unsystematically around the interior. That method contrasts with the method of Drew, who extended the border approach to produce concentric rings and thus counted in a systematic yet unproductive and novel way. Because of his border-rings approach, his purposeful attendance to covering the entire region, and his struggle to explain exactly how and where they would fit, we determined his thinking and behaviors to be consistent with the Primitive Coverer level. However, this approach represents an anomaly within our data for both research sites.

Area Unit Relater and Repeater

Like children at the Primitive Coverer level, children at the Area Unit Relater and Repeater level can also draw a complete covering but may make some errors in alignment. In contrast with children who are operating at a less sophisticated level, children thinking at the Area Unit Relater and Repeater level have a stronger constraint on object counting and the use of rows as an intuitive structure, thereby allowing a child to maintain correspondence between counting numbers and countable unit parts of a region. Children at this level also draw individual approximately equal-sized units without gaps or overlaps in addition to relating size and number of area units (i.e., recognizing that different area units will result in different measures). In Chapter 4, we described how children as young as kindergarten and continuing up through Grade 2 exhibited Area Unit Relater and Repeater thinking. Our accounts in this chapter show that thinking at the Area Unit Relater and Repeater level can persist beyond Grade 2, even into Grade 5. This finding is consistent with our theoretical perspective because hierarchical interactionism does not merely posit age-level development but rather it is nongenetic, providing an account of shifting types of thinking given educational or social experiences and a range of influences that include maturation. Thus, the persistence of levels such as Area Unit Relater and Repeater among children in Grade 5 is plausible.

For example, Anselm exhibited thinking at the Area Unit Relater and Repeater level in Grade 2 and on into Grade 5. During the initial assessment, the researcher showed him an unmarked 3×4 rectangular region and a square-inch tile. The researcher asked Anselm to draw what the rectangle would look like when he had completely covered it with squares and told him not to move or touch the square tile. He drew a complete covering of individual units, one at a time, until he had completely covered the rectangular region without gaps or overlaps. He drew a 3 × 5 array of approximately equal-sized units. Despite the alignment of units within rows in his completed drawing, Anselm's sequence of drawing actions suggested that he had not used a row as a unit of units. Because only the previously drawn adjacent units strongly guided his placement of these units and because his alignment of the anchored units was somewhat purposeful (cf. Miller, 2013), he was using the intuitive structure of a row as a marker to guide his drawing actions. We interpreted his complete covering of approximately equal-sized units and his iteration of one area unit based on rows as evidence of the Area Unit Relater and Repeater level.

Another child, Owen, also demonstrated thinking at the Area Unit Relater and Repeater level in Grades 2 to 5. We categorized Area Unit Relater and Repeater as his dominant level of thinking during Grade 2, but it was not as dominant in subsequent grades. For example, during the spring of Grade 3, the researcher gave Owen a 2 × 2 square with all four sides labeled with "2 in" and asked him to find its area. To complete this task, Owen drew units one at a time, anchoring each unit to the one above it. We noted that he did not curtail the process to produce columns of individual units but persisted in drawing individual units (see Figure 5.3).

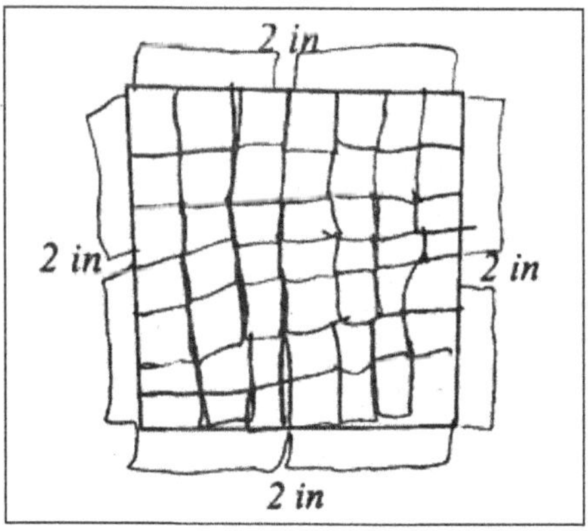

Figure 5.3. Owen exhibited drawing behaviors indicative of the Area Unit Relater and Repeater level when the researcher asked him to find the area of a 2 in. × 2 in. square.

Like Anselm, Owen created a complete covering with approximately equal-sized individual units that he imposed on the object (i.e., the squares that he drew were much smaller than a square inch). After Owen completed his drawing, he counted one unit at a time, organized by columns, to determine the number of units needed to cover the region. Although Owen restarted his count three different times, he still ended up with a miscount of "50 area squares." We saw that problem as evidence that Owen was not yet seeing a column or row as a unit of units. Instead, he showed evidence of drawing a complete covering of approximately equal-sized units and using the intuitive structure of a column to guide his drawing and counting actions. These behaviors and thinking are indicative of the Area Unit Relater and Repeater level, with the caveat that his use of the intuitive structure of a column shows a transition toward the next level, Partial Row Structurer.

Partial Row Structurer

Children thinking at the Partial Row Structurer level exhibit substantial progress in their conceptualization of planar regions as structured collections of units. They have developed spatial structuring schemes that allow them to represent individual units as rows and to accumulate units across sets of rows, understand the collinearity of rows, understand the constraint that each row must have the same number of units, and recognize a square as both a unit and as a unit of units (a row or column). At the less sophisticated Area Unit Relater and Repeater level, children draw individual units and may use rows as an intuitive structure; whereas at the Partial Row Structurer level, children typically use mixed drawing and counting actions by recognizing, drawing, and counting at least some rows (or columns) as a unit of units. In contrast with the next more sophisticated level (i.e., Area Row and Column Structurer level), children who are reasoning at the Partial Row Structurer level cannot yet attend to both a row and a column, and they struggle with the concept that the length of a line segment specifies the number of area units that will fit along that line segment.

The first time that Anselm demonstrated thinking at the Partial Row Structurer level was during the spring of Grade 2. The researcher gave Anselm two rectangular regions (7×3 and 4×6) with unnumbered rulers along a set of orthogonal sides per rectangular region and asked him which rectangular region was bigger by area.

We provided him with one square unit tile. The researcher placed it just below the 7×3 rectangular region within the first interval on the horizontal unnumbered ruler (see Figure 5.4a). The researcher asked Anselm how many unit tiles he would need to cover the whole 7×3 rectangular region and told him to try first without touching the tile, a procedure that might have allowed him to work with his hands and thus figural images of the tile. Anselm counted in sets of three (i.e., rows of three). He said,

> All three of these [dragged three fingers horizontally over three rows] which would equal 9 . . . and then plus 3 [more] plus 1 more 3 would be 12, plus another 3 would be 15, 15, and then 16, 17, 18, [pause] then 19, 20, 21. This is 21 . . . 21 squares would fit in here.

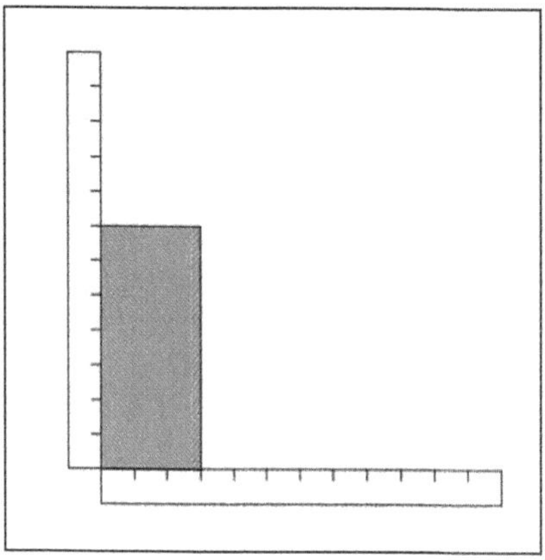

Figure 5.4a. 7 × 3 rectangular region with unnumbered rulers.

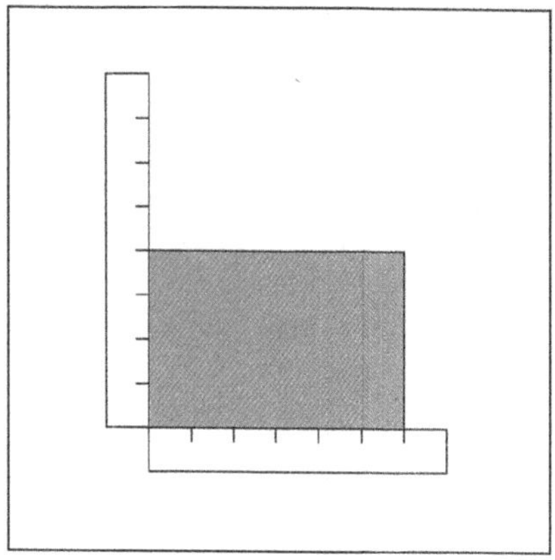

Figure 5.4b. 4 × 6 rectangular region with unnumbered rulers.

Anselm was counting rows of units before reverting back to counting individual but grouped squares (i.e., saying, "16, 17, 18, [pause] then 19, 20, and 21"). Despite his mixed counting actions, we took this process to indicate that he recognized a row as a unit of units.

Anselm used similar mixed counting actions on the 4 × 6 rectangular region (see Figure 5.4b), counting in sets of four before reverting back to counting individual but grouped squares. He said,

> 4 and 4, 8 [pointed below first and second column with two fingers], plus 2 more 12 [used two fingers to indicate his mental placement of the first and second tiles in the third column] . . . I mean it's 10 then you got 12 [used two fingers to indicate his mental placement of the third and fourth tiles in the third column], 14, 15, 16, [pointed to his mental placement of individual tiles in the fourth column, one point per count], 17, 18, 19, 20, [pointed to his mental placement of individual tiles in the fifth column, one point per count] 21, 22, 23, 24 [pointed to his mental placement of individual tiles in the sixth column, one point per count]

We also believed that Anselm's attention to a group of four squares was evidence that he was beginning to see a column of squares as a unit of units, and he appeared to understand the constraint that each row must have the same number of units. This understanding is indicative of behaviors and thinking at the Partial Row Structurer level.

Abby used dominantly Partial Row Structurer level thinking in Grades 2, 3, and 4 and continued to use thinking at the Partial Row Structurer level into Grade 5. On the initial assessment, the researcher showed Abby a gridded 4 × 5 rectangle and asked her to copy the image. Abby first drew four boundary segments to create a rectangle. Second, she drew one column of four individual units. Third,

Abby completed the first row with a total of five individual squares. Next, she curtailed her drawing actions to draw columns of individual squares by drawing three approximately parallel column segments from left to right and then two approximately parallel row segments from top to bottom. Using these mixed drawing behaviors (i.e., drawing a mixture of individual units and rows of individual units) to complete a 4×5 array is consistent with the Partial Row Structurer level.

On another initial assessment item, the researcher gave Abby a square unit tile and asked how many she would need to cover a 3×4 rectangular region. Abby iterated one area unit down three times to create the first column and then iterated it across four times to create the first row. When the researcher asked how many tiles she would need to cover the rectangle, Abby answered, "16 because 4 plus 4 is 8, and 8 plus 8 is 16." Although her numeric answer was incorrect, her thinking was quite sophisticated for a child in Grade 2. Abby did not need to iterate a single area unit across the entire rectangular region and instead curtailed her counting actions to use additive reasoning and think about groups of units through doubling. We take this reasoning to be evidence that Abby was thinking at the Partial Row Structurer level on this task.

Area Row and Column Structurer

Children who exhibit thinking at the Area Row and Column Structurer level have made a significant conceptual advance toward representing a square within an array as an individual unit, a component of a row, and a component of a column. At this level, children also show development in their coordinating schemes. They have developed the concept that the side length determines the number of area units that fit along that length in both dimensions, and they can coordinate both dimensions. That is, children thinking at the Area Row and Column Structurer level determine the number of units in a row or column, a unit of units, and iterate this unit of units repeatedly and exhaustively in one-to-one correspondence with the elements of the orthogonal column to structure an array. At the Area Row and Column Structurer level, a child may count one by one systematically (e.g., by row), apply a skip-counting scheme, or use multiplication to determine area. However, children at the Area Row and Column Structurer level may not yet be able to explain how multiplication creates a measure of area or determine area without drawing; these ideas develop at the Array Structurer level.

Anselm exhibited thinking at the Area Row and Column Structurer level in the spring of Grade 3. The researcher gave him an unmarked 7×4 rectangular region along with a ruler and asked him to find the area of the rectangle. Anselm lined up the ruler vertically, just inside the 4-inch side of the rectangular region and proceeded to iterate the ruler (using the width of a ruler as an inch) across the rectangular region, while skip counting "4, 8, 12, 16, 20, 24 . . . I would say it is about 24 squares." His action of iterating the ruler, as a representative of a unit of units (a column) of four square units across the rectangular region, furnished some evidence that he was able to think about the number of units in a row or column (i.e., a unit of units) and iterate this unit of units repeatedly and exhaustively in one-to-one correspondence with the elements of the orthogonal column to structure an array even though he missed one column of four. Although Anselm used skip counting to determine the area for this task, he structured an array by iterating a unit of units exhaustively and applied the concept that the length of a line segment determines the number of area units that fit along that length. We took this method as evidence that Anselm was operating at the Area Row and Column Structurer level.

During the fall of Grade 5, we revisited the 7 × 4 rectangle task with Anselm. This time, he measured the length and width of the rectangle as 7 inches and 4 inches and used multiplication to determine the area. He said, "28 square inches will fit inside this rectangle." When the researcher asked him to show the 28 square inches, he subdivided the rectangle with parallel row and line column segments and applied the concept that the side length determines the number of area units that fit along that length in both dimensions (see Figure 5.5). That is, he lined up the ruler with perpendicular adjacent sides of the rectangle and used the numbered tick marks on the ruler to draw parallel line segments to subdivide the figure to produce an array.

Another child, Owen, demonstrated thinking at the Area Row and Column Structurer level at the end of Grade 4. We showed him a picture of a 10 cm × 8 cm rectangular region next to a 2 cm × 3 cm gray unit, both with the linear dimensions labeled, and asked him how many gray tiles he would need to cover the rectangle exactly He said that 14 tiles would cover the rectangular region. To explain his answer, Owen drew an array of squares in the rectangular region (see Figure 5.6). He tentatively drew a column of four units, rotated 90 degrees along the 10 cm side. He next completed the row along the 8 cm top edge of the rectangular region by using two unrotated units. Thus, Owen related linear and area units along the top of the rectangular region. The researcher pointed out that the four groups of three units along the 10 cm side was not the same length as the side already marked 10. To resolve that problem, Owen crossed out one of the vertical tiles and noted that he would need 13 tiles and "one centimeter left." Although Owen did not know how to resolve the 1 cm × 2 cm region that he left uncovered, he was able to reason that the length of a line segment determines the number of area units that fit along that segment, evidence of his transition to the Area Row and Column Structurer level.

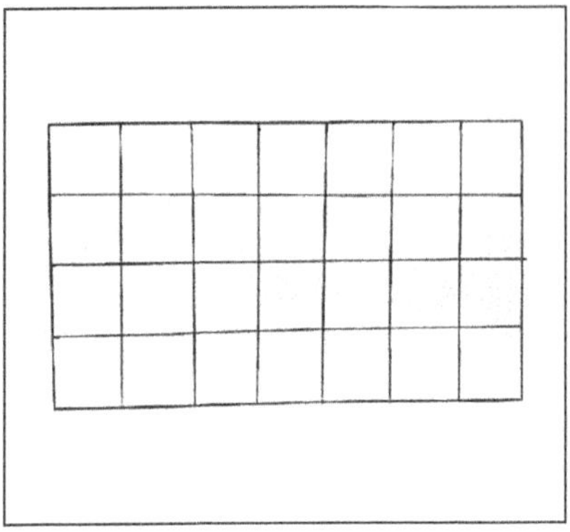

Figure 5.5. Anselm's drawing on 7 in. × 4 in. rectangular region.

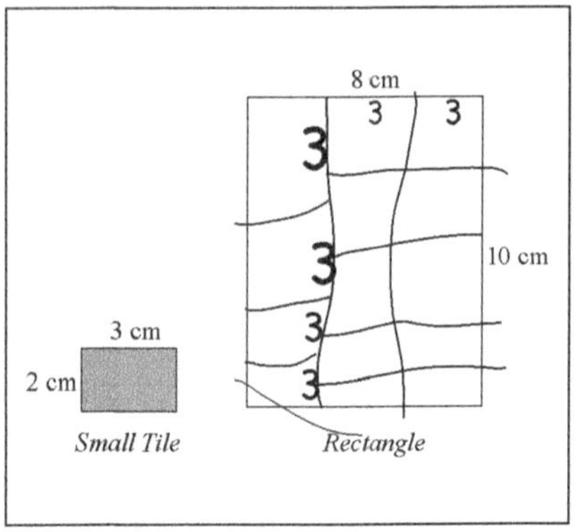

Figure 5.6. Owen's drawing to fit 2 by 3 cm tiles within an 8 cm × 10 cm rectangular region.

Area Conserver

Children thinking at the Area Conserver level are able to understand that when they decompose a region and rearrange its parts, the areas of the initial and new regions are equivalent, and area is conserved. They can reason that regions can look different but have the same area measure.

We observed great variability within and between children in their ability to conserve area. All our participants used thinking at the Area Conserver level on a minimum of five tasks during the 4-year teaching experiment. Anselm, however, consistently used thinking at the Area Conserver level throughout the teaching experiment. Anselm did not display any errors related to area conservation from the initial assessment to the final assessment, nor did he make such an error during any of our teaching episodes across 7 semesters. On 10 different tasks, he recognized the areas of two or more regions as equivalent by using visual imagery, thinking about conceptual transformations of polygons, or subdividing to unitize. For example, during Grade 4, we gave Anselm construction paper cutouts, two at a time, and asked him to compare them according to their areas. First, we presented Anselm with a square and a triangle formed by cutting a congruent square in half along its diagonal (see Figure 5.7a). Second, we presented him with a square and a rectangle formed by cutting the square half vertically (see Figure 5.7b). Finally, we gave Anselm the triangle and rectangle halves to compare (see Figure 5.7c) Anselm recognized that the triangle was half of the square and that the rectangle was half of the square. In addition, he quickly recognized that the two halves were equivalent in area because they were both halves of the same whole. (The children did not observe the creation of the triangle and rectangle halves because the researchers created them before the start of the teaching episode.)

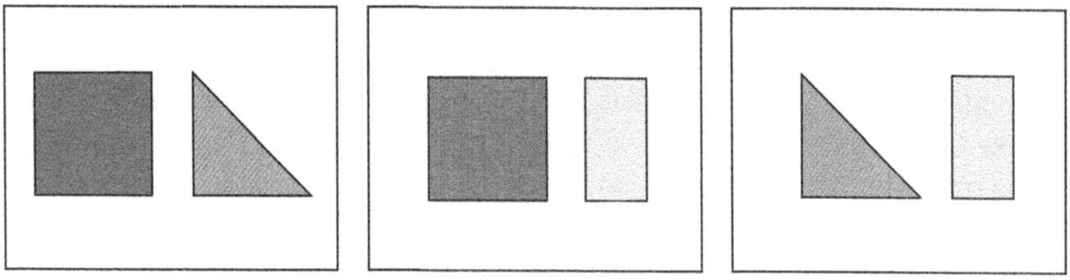

a. b. c

Figure 5.7. Square, triangle, and rectangle paper cutouts in pairs.

Another child, Owen, did not use thinking at the Area Conserver level at the beginning of the study, yet he made steady progress in the development of thinking at the Area Conserver level up into Grade 4. During Grade 2 on the initial assessment, Owen attended to only one dimension when comparing the areas of two rectangles with equal areas (see Figure 5.8). He said that even though the shapes were the "same size," the rectangle with a longer length (Shape A) covered more of the space.

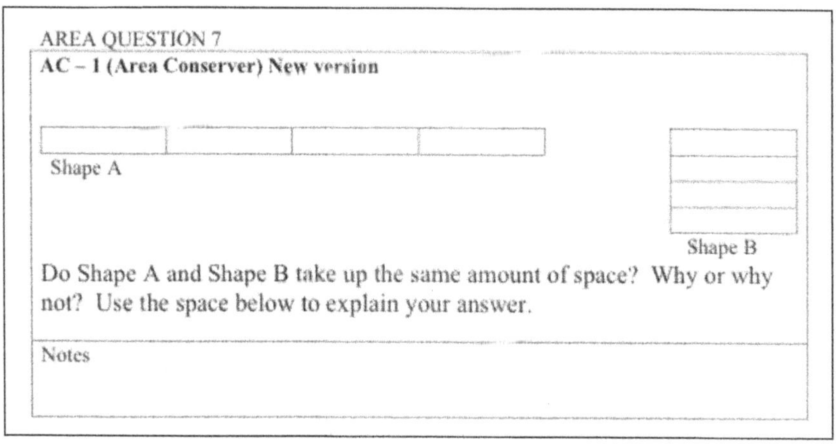

AREA QUESTION 7

AC – 1 (Area Conserver) New version

Shape A

Shape B

Do Shape A and Shape B take up the same amount of space? Why or why not? Use the space below to explain your answer.

Notes

Figure 5.8. Task posed on the initial assessment.

During the fall of Grade 4, we posed the construction-paper-cutout task to Owen (see Figure 5.7). Unlike Anselm, Owen struggled on this task. Although Owen correctly recognized that the triangle had half of the area of the square and that the rectangle had half of the area of the square, he claimed that the triangle had a larger area than the rectangle. Abby, Arielle, and Drew—in contrast to Anselm and Owen—did not exhibit consistent patterns in their thinking about area conservation; rather, they exhibited a great deal of variability on tasks requiring the conservation of area. The extreme variability of the area conservation across wide spans of development for individual children and among these children raised questions about the nature of the Area Conserver level while we considered potential improvements to the developmental progression of the LT for area measurement (see also Miller, Kara, Eames, Cullen, & Barrett, 2012). We also discussed this idea in Chapter 3 in relation to the results for the Rasch analysis of area measurement. Children's thinking about area conservation is comparatively difficult to delimit as a level in relation to other types of thinking. We will turn to this difficulty in Chapter 6.

Array Structurer

At the Array Structurer level, children can derive a measure of area by using linear measures without drawing. Although children at the Array Structurer level may be able to apply the area formula, applying the area formula alone is insufficient evidence for a claim that a child is reasoning at this level. Rather, evidence must indicate that the child can imagine, describe, and use row and column structuring; and evidence must also indicate that the child has a mental image of a spatial array.

Anselm first exhibited thinking at the Array Structurer level during the spring of Grade 4 and continued into the next year. During Grade 5, we asked Anselm to draw a rectangle that had an area of 24 square centimeters. Anselm correctly drew a 4 cm × 6 cm rectangle. When the researcher asked him to label side lengths, he labeled the sides as 6 cm, 4 cm, 6 cm, and 4 cm. We next asked him to determine the area of the rectangle that he drew. Anselm correctly answered, "24 square centimeters." When we asked him to explain his thinking, he said that he multiplied 4 times 6 to get 24. To show the 24 square centimeters, Anselm made tick marks for each centimeter along the left and bottom sides of rectangle, extended these marks to make three horizontal line segments and then five vertical line segments, and drew the correct array. Hence, he also showed evidence of having a mental image of a spatial array and an understanding that the length of a side gives the specified number of area units that fit along that segment. Therefore, we categorized Anselm as demonstrating thinking at the Array Structurer level on this task.

Arielle was the first child to exhibit thinking at the Array Structurer level in the Midwest cohort. As early as Grade 4, Arielle derived a measure of area by using linear measures without drawing, and it was her most prominent way of thinking about area measurement during Grade 5. For example, during the spring of Grade 5, the researchers asked Arielle to draw a rectangle with an area equal to $^1/_5$ the area of an unlabeled 6 × 10 rectangle. Arielle measured the length and width of the rectangle as 6 cm and 10 cm and then correctly drew a 2 cm × 6 cm rectangle. She explained that the area of the 6 cm × 10 cm rectangle is 60 and that $^1/_5$ of 60 is 12. Although Arielle said that the area of her 2 cm × 6 cm rectangle was 12, she did not identify her units. Instead of subdividing the region of the given rectangle, Arielle derived a measure of area by using linear measures without drawing and used a numerical approach to purposefully construct a rectangle that had an area of 12 square units.

Levels More Sophisticated Than Array Structurer

Although Array Structurer was the most sophisticated level described in the initial developmental progression of the LT for area measurement (see Chapter 3), our findings indicate that additional levels may be necessary to describe children's thinking during Grades 4 and 5, as well as beyond Grade 5.

During the spring of Grade 5, we asked the children at the Midwest site to use a ruler to draw a triangle with an area of 36 square centimeters. They had access to a ruler, pen, square-centimeter tiles, centimeter cubes, and a calculator. We designed this task to see whether children saw a square as a suitable unit of measure for determining the area of a triangular region and whether they would relate the area of a triangular region to the area of a rectangular region. We furnished blank paper for children to make drawings or record their own measurement process. The blank paper prompted them to draw their own images in ways that tended to reveal their internal images and concepts more than if we had given them graph paper with predrawn lattices.

Arielle was the only child in the Midwest cohort who successfully resolved this triangle production task. She drew an 8 cm × 9 cm rectangle, subdivided it into two congruent right triangles by connecting opposite corners with a diagonal, and correctly explained that each triangle has an area of 36 (see Figure 5.9a). The remaining four children—Abby, Anselm, Drew, and Owen—drew equilateral triangles, but each child thought about units in a different way that did not lead to a comprehensive solution to the problem.

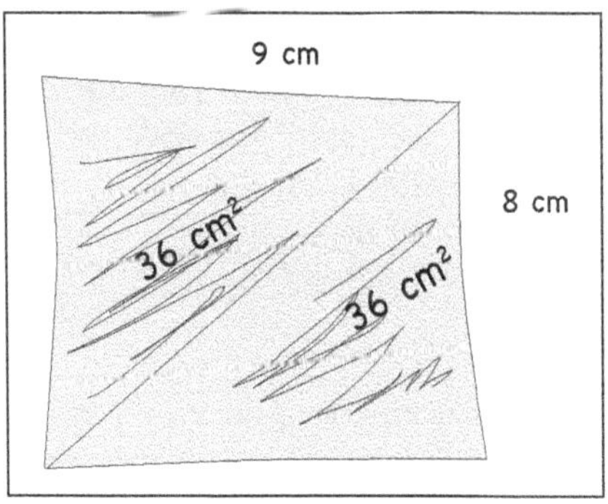

Figure 5.9a. Arielle's rectangle with an area of 36 square centimeters.

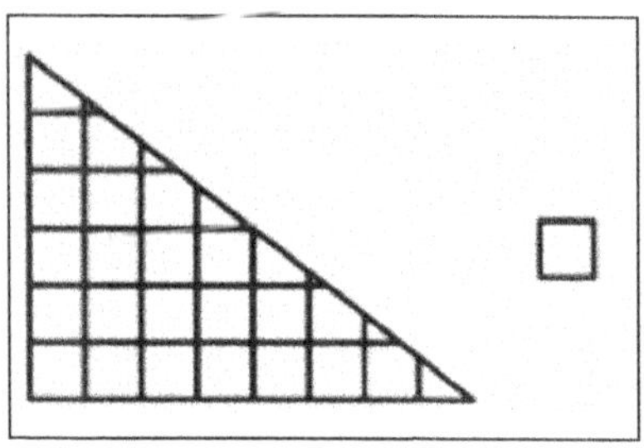

Figure 5.9b. Task posed during Grade 5.

During the same teaching episode, we asked the children at the Midwest site to compare the area of a square to the area of the triangle (see Figure 5.9b). As in the previous task, Arielle was the only child who initially thought about the right triangle as half of a rectangle. She counted eight square units in a row and six square units in a column of her imagined rectangle and multiplied 6 times 8 to arrive at an answer of 48. She then explained that "splitting" 48 into two gives the area of the triangle, which is 24. Four children—Abby, Anselm, Owen, and Drew—counted the square units one by one. Abby, Anselm, and Owen counted the whole squares and almost-whole squares each as wholes and combined partial units into wholes to add to this count. Abby counted 20 square units, Anselm counted 24, and Owen first answered 26 but changed his answer to 23½ after recounting. Drew also counted the squares one by one but considered only the 18 whole squares. When the researcher asked Drew for a careful comparison of the square unit to the triangle, he combined partial units into wholes to add to his count and said that the answer was either 20 or 21. Drew did not reject his original answer from counting units one by one. Drew's response suggests that coordinating ways of relating triangular and rectangular regions with the structuring of an array of square units in triangular regions is not trivial for some children.

On these two tasks (and others), Arielle recognized that the formula for the area of a triangle is related to the formula for the area of a rectangle, and she based her numerical operations on measures on abstracted spatial structuring. These behaviors and mental actions on objects extend beyond those of the Array Structurer level, suggesting a need for a more advanced level. We return to this conjecture in Chapter 6.

LONGITUDINAL GROWTH CHARTS

In this section, we present and discuss longitudinal growth charts that summarize the data for the five children discussed in this chapter. As previously mentioned, such charts represent only the developmental progression of an LT. These charts display our records of each child's development in area measurement through the levels of the LT. The charts are chronological records; reading them from left to right shows the levels corresponding to each school year from Grade 2 through Grade 5. The lightly shaded vertical regions denote the summers separating the school years. As previously explained in Chapters 2 and 4, the thin vertical line segments represent the levels of thinking demonstrated by the child during a given teaching episode, and the vertical rectangles represent varying levels of thinking (see Chapter 2, page 45, for a more detailed description of the growth-chart components).

Abby's Growth

Abby demonstrated thinking at the Partial Row Structurer level during the spring of Grade 2 (see Figure 5.10). Although she was using thinking at a mixture of the Partial Row Structurer, Area Row and Column Structurer, and Array Structurer levels by the end of the study in Grade 5, she continued to exhibit thinking at the Partial Row Structurer level for nearly 30 months, a longer period of time than any other child in the cohort. Although we observed behaviors that suggest that she was thinking carefully about rows and columns of units during Grade 4, she struggled in the latter portion of Grade 4 and into Grade 5 with the concept that the length of a line segment specifies the number of area units that will fit along that line segment. Even with instruction that focused on the use of square and nonsquare units to emphasize the collections of units in each row (for more information on this instruction, see Kara, Eames, Miller, Cullen, & Barrett, 2011), Abby did not explain the use of linear measures along rectangles as a way of finding the number of units in rows and the number of rows necessary to fill a rectangular region. We took this outcome as an indication that she might have been unable to relate linear measures to area measures conceptually even though she used the procedure.

This prolonged use of the same level of thinking was cause for further investigation into her development of related concepts and skills, as well as other possible influences, which might include textbook treatments of the topic or patterns of instruction in her classroom. Our retrospective analysis indicated that neither tasks posed by Abby's classroom teachers nor those presented in her textbook required the use of the concept that the length of the line segments along perpendicular sides of a rectangle specifies the number of area units needed to form a row along one of those sides. We believe that more extensive and consistent instructional experience with area-covering tasks coordinated with length-measurement tasks would have supported her development of subsequent levels of thinking.

During a teaching episode during the spring of Grade 4, Abby demonstrated a brief pattern of thinking at the Primitive Coverer level, a level that she had not exhibited in the teaching experiment since the spring of Grade 2. This fall back to the Primitive Coverer level may have been caused by the unique demands of a task involving measuring areas of nonrectilinear figures (see Chapter 6). Because of her subsequent performances at more sophisticated levels, we did not take it to indicate a backward developmental change in thinking level. We believe that thinking at the Primitive Coverer level was adequate to meet the particular task. Throughout the study, we designed and posed sequences of tasks, which children at multiple adjacent levels could approach, with extensions to elicit thinking at more sophisticated levels and interventions to scaffold growth for children at less sophisticated levels.

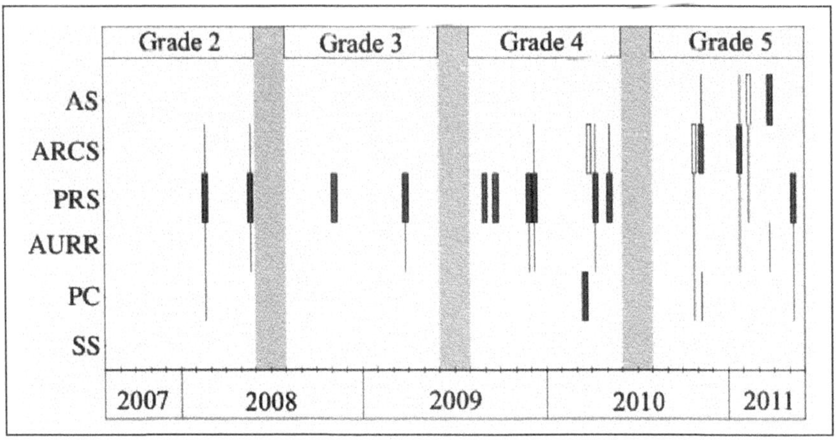

Figure 5.10. Abby's longitudinal growth chart. SS = Side-to-Side Area Measurer, PC = Primitive Coverer, AURR = Area Unit Relater and Repeater, PRS = Partial Row Structurer, ARCS = Area Row and Column Structurer, and AS = Array Structurer.

Anselm's Growth

Anselm exhibited thinking at the Area Unit Relater and Repeater level on the initial assessment but transitioned to thinking at the Partial Row Structurer level by the end of Grade 2. He also displayed another shift to thinking at the Area Row and Column Structurer level starting in the fall of Grade 4, with a fall back to thinking at the Primitive Coverer level in late fall of Grade 4. In contrast with Abby's fall back to Primitive Coverer during the spring of Grade 4 when she was measuring areas of nonrectilinear figures, we observed Anselm's fallback in two consecutive interviews, during which Anselm confounded perimeter and area and struggled to count border units correctly, both for rectangular regions. Our retrospective analysis indicates that in length contexts, Anselm struggled with whether to count the intervals or the tick marks as the length unit during Grade 3 and on into Grade 4. We conjecture that this problem affected his thinking about area measurement during part of his Grade 3 experience. This phenomenon is consistent with our theoretical perspective of hierarchical interactionalism and the tenet that concepts and skills develop in tandem (e.g., the concept of length-unit iteration develops in tandem with the concept of area-unit iteration). Furthermore, when multiple concepts interact, the development of one concept can work for or against the development of another (cf. Baroody, Lai, & Mix, 2006; Greeno, Riley, & Gelman, 1984; Sarama & Clements, 2009b). Although Anselm appeared to use thinking at the Array Structurer level during the last 2 years of the study, this transition did not seem complete before the end of Grade 5.

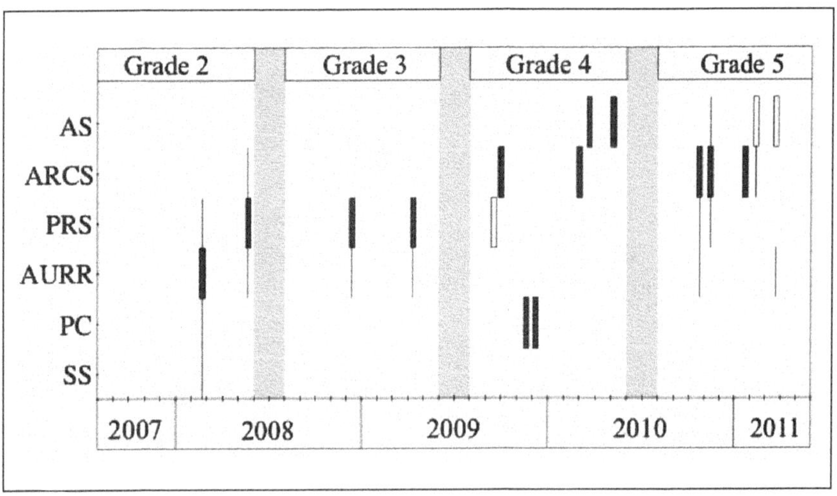

Figure 5.11. Anselm's longitudinal growth chart. SS = Side-to-Side Area Measurer, PC = Primitive Coverer, AURR = Area Unit Relater and Repeater, PRS = Partial Row Structurer, ARCS = Area Row and Column Structurer, and AS = Array Structurer.

Arielle's Growth

During the spring of Grade 2, Arielle exhibited thinking at the Partial Row Structurer level. She seemed to make a clear shift to thinking at the Area Row and Column Structurer level by the spring of Grade 4, shifting to thinking at the Array Structurer level by the fall of Grade 5 (see Figure 5.12). In contrast to the other children, Arielle fell back to thinking at the Partial Row Structurer level during the spring of Grade 4 before returning to dominantly Area Row and Column Structurer level thinking during the fall of the next year. On the novel tasks used during that time (e.g., see Figures 5.6 and 5.9b), Arielle struggled with the concept that the length of a line segment specifies the number of area units that will fit along that line segment. Hence, she illustrated a fallback in her thinking to use a less advanced approach to handle a more complex task. In general, however, her work exemplified a strong upward transition into thinking at the Array Structurer level for Grade 5.

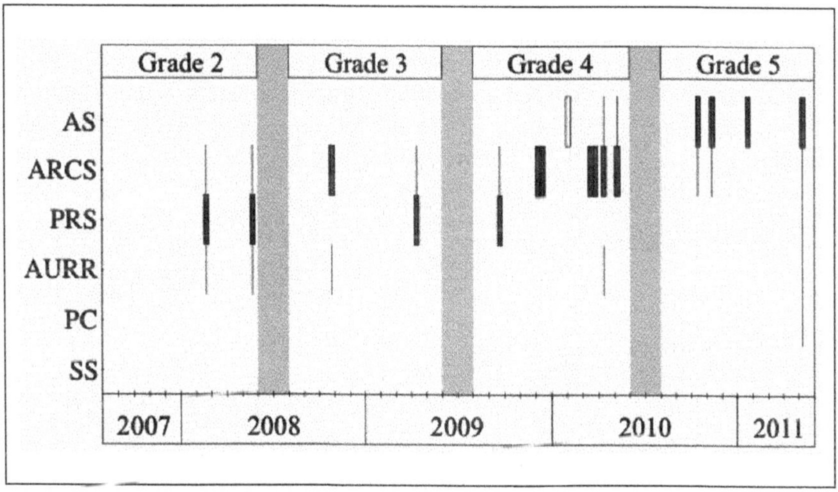

Figure 5.12. Arielle's longitudinal growth chart. SS = Side-to-Side Area Measurer, PC = Primitive Coverer, AURR = Area Unit Relater and Repeater, PRS = Partial Row Structurer, ARCS = Area Row and Column Structurer, and AS = Array Structurer.

Drew's Growth

Drew exhibited thinking at the Primitive Coverer level during the spring of Grade 2 and the fall of Grade 3 (see Figure 5.13). Although he started to show signs of using thinking at the Partial Row Structurer level as early as the spring of Grade 3, he did not completely transition from using thinking at the Primitive Coverer level until the spring of Grade 4. He transitioned through the Area Unit

Relater and Repeater and Partial Row Structurer levels and into thinking at the Area Row and Column Structurer level by the fall of Grade 5.

Although Drew did not display Area Unit Relater and Repeater as his dominant level of thinking during any of our teaching episodes, he did demonstrate thinking at the Area Unit Relater and Repeater level during the study. Because of the nature of our methods, we are not able to claim that he did not exhibit Area Unit Relater and Repeater thinking as his dominant level during the time between Grades 2 and 5—he may have done so in our absence. Such determinations would require a more intensive collection of data over the span of development, most likely during Grade 3. Moreover, we may not have provided a task that required more than a Primitive Coverer level but less than a Partial Row Structurer level of thinking and behaviors.

We characterized Drew's dominant level of thinking as Primitive Coverer for two separate teaching episodes during the fall of Grade 4; this level represented an apparent fallback. Drew oscillated between dominantly Primitive Coverer level thinking and Partial Row Structurer level thinking in that timeframe. This phenomenon demonstrates the variable nature of scheme changes for individual children, as described by our theoretical perspective of hierarchical interactionalism. Despite his oscillation between levels during the fall of Grade 4, Drew showed advancement beyond thinking at the Partial Row Structurer level to thinking at the Area Row and Column Structurer level during the spring of Grade 4, as well as during Grade 5.

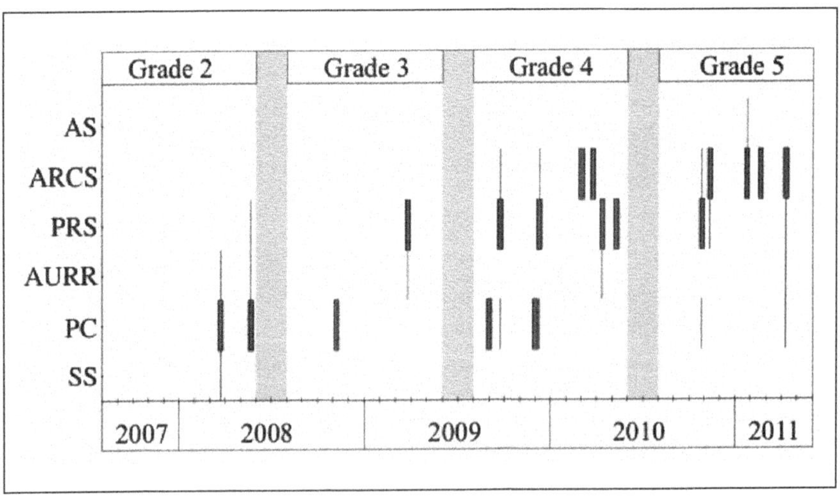

Figure 5.13. Drew's longitudinal growth chart. SS = Side-to-Side Area Measurer, PC = Primitive Coverer, AURR = Area Unit Relater and Repeater, PRS = Partial Row Structurer, ARCS = Area Row and Column Structurer, and AS = Array Structurer.

Owen's Growth

As shown in Figure 5.14, Owen exhibited thinking at the Area Unit Relater and Repeater level on the initial assessment and eventually demonstrated thinking at a mixture of the Area Row and Column Structurer and the Array Structurer levels by the end of the study. Owen displayed two main shifts in his thinking. In the spring of Grade 3, he shifted from thinking at the Area Unit Relater and Repeater level to thinking at the Partial Row Structurer level. In the fall of Grade 4, Owen shifted from using Partial Row Structurer level thinking to Area Row and Column Structurer level thinking. Like Anselm, Owen displayed signs of thinking at the Array Structurer level, but this transition did not seem complete before the end of Grade 5. In contrast to the other focus children, Owen rarely fell back from his dominant level of thinking before Grade 5.

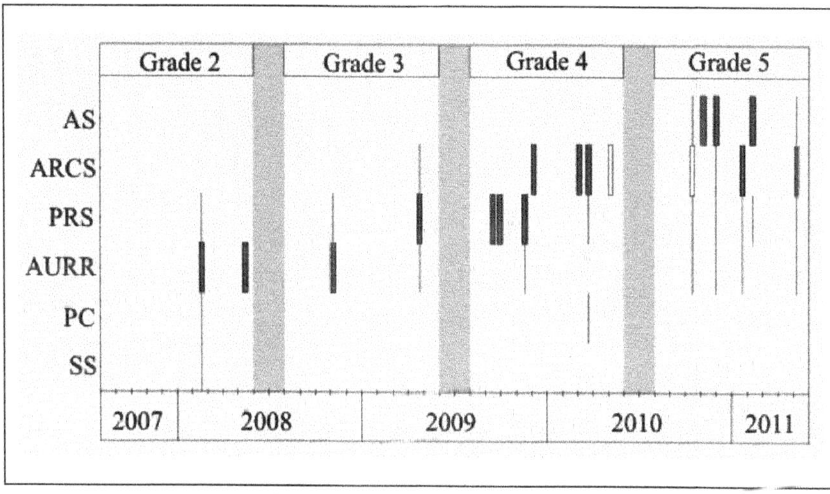

Figure 5.14. Owen's longitudinal growth chart. SS = Side-to-Side Area Measurer, PC = Primitive Coverer, AURR = Area Unit Relater and Repeater, PRS = Partial Row Structurer, ARCS = Area Row and Column Structurer, and AS = Array Structurer.

The overall growth patterns during 3.5 school years, from spring of Grade 2 through spring of Grade 5, illustrate that these five children advanced one, two, or three LT levels. One naturally wonders about the differences in the timing and shape of the growth trends that each child exhibited. Arielle and Owen demonstrated the most consistently upward trend, showing increasing sophistication in thinking over time. The other three children displayed progress, but their development appeared more protracted, with several fallbacks to previous levels along the way. We take this outcome as evidence that supports our claim of leveled progress, with two caveats: (a) that individual children may not attain the immediate level that others in their age group and age attain and (b) that individual children may exhibit variable progress that lasts as long as a year (e.g., Abby's plateau at the Partial Row Structurer level for 30 months).

DISCUSSION OF LONGITUDINAL GROWTH CHARTS

Figure 5.15 is a composite view of all five children, produced by overlaying the growth charts of Abby, Anselm, Arielle, Drew, and Owen. This chart indicates some positive growth through the LT for area measurement while the children moved from Grade 2 through Grade 5, progressing from Area Unit Relater and Repeater in Grade 2 up through Array Structurer by Grade 5.

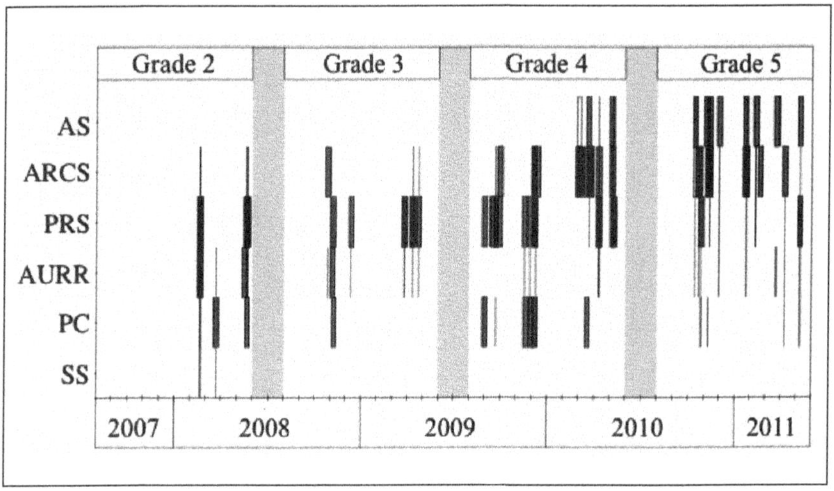

Figure 5.15. Composite longitudinal growth chart. SS = Side-to-Side Area Measurer, PC = Primitive Coverer, AURR = Area Unit Relater and Repeater, PRS = Partial Row Structurer, ARCS = Area Row and Column Structurer, and AS = Array Structurer.

For the children in the Midwest cohort, less observable growth occurred in Grades 2 and 3 than in the later grades, possibly because of a limited set of data points to reveal such growth. Fewer teaching episodes focused on area measurement during Grades 2 and 3, and classroom instruction in those grades included less curricular attention to area concepts. By the end of Grade 3, the dominant level of thinking for all five children was Partial Row Structurer. During Grade 4, the Midwest research team conducted many more interviews on area than they had for Grades 2 and 3, and classroom instruction of the Midwest cohort began to emphasize the use of arrays and measures of area, as well as formulas for computing area. The growth trends on our charts reveal movement through the LT levels for Anselm, Arielle, Drew, and Owen during that year. Although Abby's chart shows that we characterized her dominant level of thinking as Partial Row Structurer throughout Grade 4, she demonstrated growth by showing some evidence of thinking at the Area Row and Column Structurer level by the end of that year.

All the children demonstrated two or three different dominant levels of thinking during Grade 4. Abby, Arielle, and Owen demonstrated two; and Anselm and Drew demonstrated three. As a collection, these children exhibited five levels of thinking, spanning Primitive Coverer to Array Structurer, as Figure 5.15 indicates. Variation may have existed in cognitive demands from the interview tasks, allowing multiple approaches to resolve questions; this variation could account for sustained use of lower level thinking during Grade 4. By Grade 5, the children had stabilized at either Area Row and Column Structurer or Array Structurer. Much less variability is evident during this grade, with the data points for all the children except Abby clustered either at the Area Row and Column Structurer or the Array Structurer levels.

Several patterns emerge when we focus on specific LT levels throughout the 4-year span. The initial level designated in Figures 5.10 to 5.15 is Side-to-Side Area Measurer. Anselm, Drew, and Owen exhibited this level during Grade 2 but not as their dominant level of thinking. Thinking at the Side-to-Side Area Measurer level did not appear during Grades 3, 4, and 5, most likely because of the lack of activities involving complete sets of square tiles that children might use to cover a region in a way that would lend itself to that level.

None of the children in the Midwest cohort exhibited Array Structurer, the most sophisticated level displayed in these figures, until the end of Grade 4. These children rarely encountered examples of rectangular regions subdivided into arrays before Grade 3, and they would have had little opportunity to work with arrays as models for additive operations, the major curricular focus in mathematics during Grade 2. Four of the five children—Abby, Anselm, Arielle, and Owen—displayed the Array Structurer level as their dominant level of thinking by the end of Grade 5.

The levels between Side-to-Side Area Measurer and Array Structurer (i.e., Primitive Coverer, Area Unit Relater and Repeater, Partial Row Structurer, and Area Row and Column Structurer) appeared as dominant levels of thinking for the children in the Midwest cohort during this 4-year study, with each of the children showing slightly different trends. Thinking at the Primitive Coverer level was dominant for three of the children—Abby, Anselm, and Drew—for at least one teaching episode during the study. For Abby and Anselm, this level occurred during Grade 4 for different reasons, as discussed previously. Although we do have evidence of children in the Midwest cohort exhibiting thinking at the Primitive Coverer level, we did not observe any of them transition in and out of that level, as we observed with the children in the Northeast cohort. This outcome may be attributable to the broad range of existing knowledge and levels of thinking among the children in the Midwest cohort.

Thinking at the Area Unit Relater and Repeater level was present throughout the 4-year study, but it was rarely the dominant level of thinking for the focus children. During the initial assessment, Anselm used thinking at the Area Unit Relater and Repeater level, but he quickly transitioned to higher level thinking after that assessment. Owen was the only other child to use dominantly Area Unit Relater and Repeater level thinking during the study, and he used it only during Grade 2 and the fall of Grade 3. Because we started working with the children in the Midwest cohort during Grade 2, two of our children—Abby and Arielle—had already moved beyond that level. We also note that our grade-related participant selection criteria may have limited our opportunity to observe that level of thinking.

The Partial Row Structurer level of thinking was the most prevalent level in our study of children in the Midwest cohort. All five children exhibited this level of thinking in multiple grades, especially in Grades 3 and 4. These children tended to plateau at this level for as few as 12 months to as many as 30 months, indicating that this transition took time and required thinking that was qualitatively different from both the previous level and the following level. We also attribute this variability in the duration of thinking at the Partial Row Structurer level to our use of tasks during Grades 4 and 5 that allowed such reasoning without requiring thinking at a more advanced level. With one exception, thinking at the Area Row and Column Structurer level was dominant for children only during Grade 4 or Grade 5.

We posit an increasing depth of cognitive resources for children while they progress to higher levels in the developmental progression, which results in greater variation in the thinking displayed. For example, a child operating at a more sophisticated level (e.g., level n) can reach back to thinking at a less sophisticated level (e.g., level $n - 1$, $n - 2$) and likely will reach back to the least sophisticated level that will allow him or her to solve the task correctly and more simply, especially if the task does not require a higher level or if using more sophisticated thinking will not save time or mental effort (e.g., the persistence of thinking at the Partial Row Structurer level into Grades 4 and 5 for some children, as described in the previous paragraph).

A child operating at a less sophisticated level does not have this same flexibility because he or she cannot choose to reach forward to a more sophisticated level. We also conjecture that a child may fall back to a previous level when presented with a task that requires thinking from a level beyond the child's current level (cf. Kieren & Pirie, 1992). In these situations, a child may be struggling to solve the task and thus may fall back to a less sophisticated level at which he or she will have more confidence in his or her thinking, as Arielle did during the spring of Grade 4. These reachbacks, fallbacks, and periodic growth spurts characterize a child's growth through levels according to our theoretical perspective of hierarchic interactionalism (Clements & Sarama, 2007).

Finally, the composite growth chart shown in Figure 5.15 has implications for classroom teachers who address the teaching of area or who use area measure as a model for multiplication. One such implication is the potential need to differentiate instruction according to the dominant levels of thinking for a group of children. The five children in the Midwest cohort exhibited key growth phases in area measurement at different times in Grade 4, demonstrating Primitive Coverer, Partial Row Structurer, Area Row and Column Structurer, and Array Structurer levels of thinking during that year of our study. This outcome suggests that at least within Grade 4, teachers can expect instructional challenges and opportunities because of the substantial individual variability among these four levels. Therefore, our findings indicate a potential need for teachers to prepare a wide variety of area-measurement tasks that are both accessible to children who are operating at a variety of adjacent levels and suitable for supporting children's growth from one level into the next.

Chapter 6

Integration of Results: A Revised Learning Trajectory for Area Measurement

Jeffrey E. Barrett
Illinois State University
Douglas H. Clements
University of Denver
Julie Sarama
University of Denver
Amanda L. Miller
Illinois State University
Craig J. Cullen
Illinois State University
Douglas W. Van Dine
Metropolitan State University of Denver
Katherine Newburgh
University of Denver
Maria Vukovich
The Center for Victims of Torture
Cheryl L. Eames
Southern Illinois University, Edwardsville
David Klanderman
Trinity Christian College

Prior to this longitudinal study, extant research indicated that four main schemes supported the development of area measurement: covering, quantifying, subdividing, and spatial structuring (see Chapter 3) based on covering by tiling or drawing tasks. During this study, we extended this research by posing a broader range of tasks: We included tasks for comparing regions and structured groups of units, producing regions of specific quantities, and iterating units and collections of units. We asked children to think about square units within rectangular regions, as well as nonsquare but rectilinear units within rectilinear regions and square units within curvilinear regions. We developed these tasks by attending to a wider set of principles of instruction related to measurement (e.g., Battista, 2012; Chappell & Thompson, 1999a, 1999b; Newcombe & Huttenlocher, 2000; van den Heuvel-Panhuizen & Buys, 2008). Developing a set of complementary tasks contributed to our understanding of how children resolved a wider variety of area-measurement tasks, allowing for the expansion of our theoretical account of children's mental actions on objects (i.e., the second column in an LT).

In our retrospective analysis of this study, we compared the developmental progression for area measurement with the developmental progressions for length and volume measurement. We noticed some similarities in how children built and manipulated first individual units, then a unit of units, and eventually a unit of units of units, yet these behaviors and mental actions were not yet explicit across the levels of the three LTs. Hence, we sought to elucidate these behaviors and mental actions on objects across all three revised LTs.

HYPOTHESIZING A NEW DEVELOPMENTAL PROGRESSION FOR AREA MEASUREMENT

This chapter reviews the initial developmental progression of the LT for area measurement by moving from level to level while integrating qualitative and quantitative analyses from Chapters 3, 4, and 5. We review the evidence to either confirm or disconfirm the developmental progression across LT levels, and we discuss modifications. We conclude this chapter by describing the new developmental progression of the LT for area measurement.

Incorporating Pre-Area Quantity Recognizer

The first level of the initial developmental progression of the LT for area measurement was Pre-Area Quantity Recognizer. Our data, as reported in Chapter 4, supported an initial level in which children perceive space and objects within space. However, we determined that the label *Pre-Area Quantity Recognizer* was inadequate for describing this level. Because this level was the one at which children recognized area as an attribute, we believed that Area Quantity Recognizer was a more apt name for this level (cf. Length Quantity Recognizer level of the LT for length measurement and Volume Quantity Recognizer level of the LT for volume measurement). We have incorporated aspects of the former Pre-Area Quantity Recognizer level into the new Area Quantity Recognizer level, whereas we removed other aspects from the LT. Results from both quantitative and qualitative analysis confirmed that children in pre-K can identify area as an attribute; however, they initially show little understanding of the concept of area and draw "mostly closed shapes and lines with no indication of covering the specific region" (Sarama & Clements, 2009b, p. 300). The boundary of the region, however, does seem to constrain these drawings (cf. Mulligan, Prescott, Mitchelmore, & Outhred, 2005).

Incorporating Area Simple Comparer

Our results also confirmed that children in pre-K may use side-matching strategies to compare areas (cf. Silverman, York, & Zuidema, 1984); we considered these behaviors as part of the Area Simple Comparer level in the initial LT for area measurement. We saw these behaviors occurring in conjunction with drawings, as described previously; and we therefore incorporated them into our new Area Quantity Recognizer level. In addition, the initial Area Simple Comparer level focused on how children operating at this level compare areas; however, it did not account for how children operating at this level draw, tile, or build regions of area. On the basis of our work in the present study, we saw that during tiling tasks children placed tiles on or in the region without producing a complete tiling, instead leaving gaps or overlaps (see Area Simple Comparer section of Chapter 4).

Adding Area Quantity Recognizer

The new level of Area Quantity Recognizer incorporates aspects of the first two levels from the initial developmental progression for area measurement, and it parallels the developmental progressions for length and volume measurement. A child at the Area Quantity Recognizer level perceives two-dimensional space and objects within that space. On tiling tasks, a child at the Area Quantity Recognizer level may leave large gaps between tiles or allow tiles to overlap one another or the boundary. On drawing tasks, the child may unsystematically draw approximations of circles or other figures but attend to the constraints of the bounded region. On both tiling and drawing tasks, a child at this level shows no indication of attempting to cover the region when tiling. In addition, when a researcher asks a child at the Area Quantity Recognizer level to compare regions, he or she may use side-matching strategies because the salience of linear quantity interferes with his or her attention to area and its quantification.

Redefining Side-to-Side Area Measurer as Physical Coverer and Counter

Children operating at this level are in the process of developing or have developed the ability to tile a rectilinear region completely when the tiles easily cover it without a need for rotating the tiles. We saw children progress to this level shortly after solidifying thinking at the Area Quantity Recognizer level. This finding provided further evidence supporting the placement of the Side-to-Side Area Measurer level. Although these children draw approximately rectangular shapes within the region, they often leave gaps or overlaps or align only in one dimension. When children at this level count to find a total after drawing or covering, they rely on existing guides to direct their counting; hence, they count one by one. When they move beyond limited side-matching strategies in comparing two regions, children at this level may superimpose one region on the other by placing one region or rectangle on top of the other one.

After revising this level to consider children's thinking in the contexts of drawing, building, and counting, we renamed it Physical Coverer and Counter. This new title reflects that this level is the first one at which children may create a complete covering; however, they accomplish this task only with physical units (e.g., foam squares or square plastic tiles). Children at the Physical Coverer and Counter level do not yet draw a complete covering.

Renaming Primitive Coverer as Complete Coverer and Counter

Our results from the Northeast cohort (see Chapter 4) provided confirmation for the Primitive Coverer level from our initial developmental progression. Because of those results, we renamed it the Complete Coverer and Counter level and placed it after the level that we now call the Physical Coverer and Counter level. At this level, children progress from the ability to create complete physical coverings to drawing a complete covering without gaps or overlaps. However, children still may make errors of alignment, and they do not yet recognize the need for equal-sized units, partially because they have not yet developed the concept of an area unit. When a researcher asks children at this level to cover without a complete set of physical tiles, they may begin to iterate but are unable to maintain equal iterations. After children cover by iterating, drawing, or tiling when counting to a total, they may be unsystematic in counting individual shapes. Some children may first count around the boundary and then unsystematically in the interior by losing track of what they have counted, miscounting some shapes, and double counting others (see Edith's counting illustrated in Figure 4.10a in Chapter 4 and Anselm's counting illustrated in Figure 5.1 in Chapter 5).

This level is also the first one at which a child can produce a region of a specified area when a researcher furnishes physical tiles, as long as the child has more than enough tiles to produce a region with that area. For example, when a researcher gave children more than 12 foam squares, they could build a rectangle with an area of 12 squares. We renamed this level Complete Coverer and Counter to indicate an increasingly comprehensive way of thinking about covering a region.

Revisiting Area Unit Relater and Repeater

Our results confirmed the need for the Area Unit Relater and Repeater level and its placement within the LT. At this level, children develop the concept of a unit of area, and it is the first level at which children can successfully iterate an area unit. Thus, they draw equal-sized units one at a time in addition to counting individual units. Furthermore, previously drawn or adjacent units strongly guide their placement of units. Children at this level tend to use the intuitive structure of a row to guide their drawing and counting actions (see the discussion of Edith's repeating of a physical tile in the Area Unit Relater and Repeater section in Chapter 4 and Owen's drawing in Figure 5.6 in Chapter 5).

As we have argued, this level is the first one at which children can iterate an area unit successfully. In addition, children can make a drawing of a region of specified area by using tiles, even if they do not have enough tiles to fill the region. For example, when a researcher gives children at the Area Unit Relater and Repeater level fewer than 12 foam squares, they can use iteration to construct a rectangle with an area of 12 square units. They are also capable of relating the size and number of area units needed to cover a region and recognize that different-sized area units will result in different measures.

Expanding Partial Row Structurer

As noted in Chapter 5, the five children in the Midwest cohort exhibited thinking at the Partial Row Structurer level over an extended period; it was their dominant level of thinking for at least 12 months and for as long as 30 months before they shifted to the next level. After a retrospective analysis of the results, we determined that this transition from Area Unit Relater and Repeater, through Partial Row Structurer, and into Area Row and Column Structurer is more complex than previously described. We argue that two distinct sublevels comprise the transition from Area Unit Relater and Repeater into Area Row and Column Structurer. We present our argument in two parts.

First, children at the Partial Row Structurer level develop the capacity to build and manipulate a unit of units, but this construction does not necessarily involve entire rows or columns. According to the initial developmental progression of the LT, children may draw individual units one by one and then curtail these actions to draw rows or columns of individual units, or they may draw rows or columns of individual units but then revert to drawing individual units one by one. Yet, our results indicated that rows and columns are not the only units of units that these children may build and manipulate. They may also group units into portions of rows or columns. For example, at the Northeast site in a teaching episode during the fall of Grade 2, Edith counted six tiles drawn across the top and down the left side of a 3 × 4 rectangular region. She then iterated in the empty space to determine another group of six, which she combined with the first to obtain a total area of 12. This distinction was absent in the Partial Row Structurer level of the initial developmental progression.

Second, although children at the Partial Row Structurer level initially ignored number labels, tick marks, dots, and other displays of linear dimensions along sides, they began to recognize the usefulness of these displays of linear dimensions before they developed clearly coordinated models of rows and columns as a basis for measuring area (i.e., operating at the Area Row and Column Structurer level). That is, children at the Partial Row Structurer level began to use the length or the width to constrain the unit size. They coordinated linear and area units, but only in one dimension, not in both dimensions. For example, they might coordinate the height of the unit with the height of the entire region. However, they could not yet coordinate linear and area units in both width and height, which is a characteristic of thinking at the next level, the Area Row and Column Structurer level.

These results prompted us to make two changes. First, we renamed the Partial Row Structurer level. It became the Initial Composite Structurer level. This new title highlights that it is the first level at which children can build and manipulate a unit of units (a composite unit) when covering (by drawing or iterating), producing, or comparing regions. Second, because our results suggest that a scheme for operating on a unit of units develops independently with respect to the development of a scheme for coordinating and relating dimensions, we identified potential sublevels. We use Sublevel A to describe the use of a unit of units and Sublevel B to describe the initial coordination of linear and area units in one dimension. More research is necessary before we can decide whether we need to separate these two sublevels into two distinct levels and to confirm any sequential claim for the two sublevels. That is, the question remains: Does one sublevel precede the other, or are they concurrent separate components of the Initial Composite Structurer level?

Revisiting Area Row and Column Structurer

Children in the Midwest cohort furnished evidence of the Area Row and Column Structurer level during Grades 4 and 5. We confirmed that when researchers provide children at the Area Row and Column Structurer level with tasks designed to investigate thinking and behaviors at that level, the children use available information about dimensions to coordinate linear measurements, relate linear and area units, construct arrays with parallel row and column segments, and connect the interpretation of multiplication as repeated addition with the notion of an iterated row of identical area units.

Because the children in the Midwest cohort exhibited thinking at the Partial Row Structurer level for an extended period, we were able to investigate the transition from thinking at the Partial Row Structurer level to thinking at the Area Row and Column Structurer level in ways not previously reported. To investigate this transition, we posed area-measurement tasks consisting of comparisons between nonsquare rectilinear units and rectangular regions. Similarly, we posed tasks to compare square units with rectilinear but nonrectangular regions.

For example, in the fall of Grade 5, we gave children at the Midwest site a rectangular region and a right scalene triangular unit. Number labels along each segment represented the linear units on both the region and the unit. Children achieved varying levels of success. Abby used the span between her finger and thumb (cf. Cullen & Barrett, 2010) to transfer the length of the hypotenuse of the triangular unit and created a complete covering by drawing individual approximately equal-sized triangular units within the rectangular region. She aligned rectangular composite units consisting of two triangular units into two rows and two columns by first drawing one triangular unit in the top-left corner. She then drew a row segment before completing a second triangular unit next to her first, thereby creating a rectangular composite unit from the two triangles. Next, she drew in a diagonal to create third and fourth triangular units within the second rectangular composite unit in the upper-right-hand corner of the region, stating "And that's half the snowflake." Abby understood the constraint that each row must have the same number of units. When the researcher prompted her to finish her drawing, Abby extended the column segment from the bottom left of the first rectangular composite unit to the bottom boundary before drawing in two diagonals to create the fifth, sixth, seventh, and eighth triangular units (see Figure 6.1).

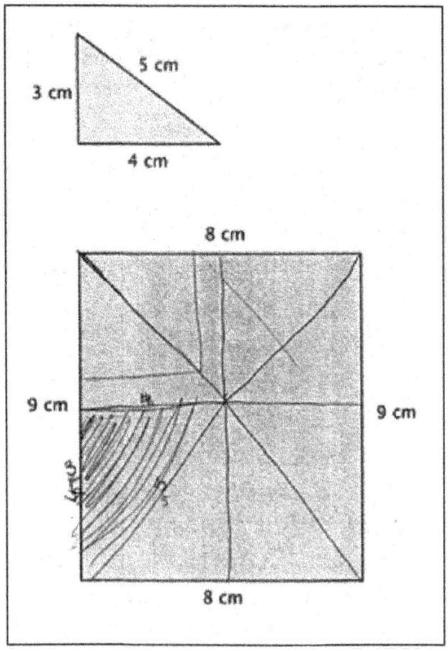

Figure 6.1. Abby's drawing on the triangular unit task.

Characterizing Abby's thinking with the initial area LT was difficult because she was operating on a right scalene triangular unit rather than on a square unit. We conjecture that covering with a triangular unit is harder than covering with a square unit because a square unit has more rotation and reflection symmetry. The initial developmental progression for area did not address nonsquare units. To characterize Abby's thinking with respect to the initial developmental progression, we generalized her operations on units and deemphasized her attention to the set of triangular shapes. By drawing a complete covering of individual triangular units of approximately equal size, Abby demonstrated thinking at least at the Area Unit Relater and Repeater level. Because her complete covering consisted of a mixed strategy, drawing individual triangular units within rows before curtailing her actions to draw rows (and columns) of individual rectangles, Abby also demonstrated thinking at the Partial Row Structurer level.

We did not classify these mental and physical actions on objects at the Area Row and Column Structurer level because Abby did not completely integrate her global and local actions. By one interpretation, Abby did not use global structuring by drawing only parallel row and line column segments, nor did she use the length of a side to determine the number of area units that fit along that length. Hence, she did not yet demonstrate thinking at the Area Row and Column Structurer level. By another interpretation, the task did not lend itself to an efficient grouping of triangular units along a single side; rather, pairs of such triangles functioned as a composite unit along a side.

Abby's thinking on this task was consistent with her thinking on other nonsquare unit tasks, and it was parallel with the thinking of the other children in the Midwest cohort by this point in the longitudinal study. By asking children to think about square units within rectangular regions, as well as nonsquare but rectilinear units within rectilinear regions, we increased our understanding of how these children resolved a variety of area-measurement tasks (see also Kara, Eames, Miller, Cullen, & Barrett, 2011, for more information on nonsquare unit tasks). We modified levels throughout the developmental progression for area measurement, including Area Row and Column Structurer, to delimit behaviors and mental actions on objects based on unit or region shape.

On the basis of our work with nonrectilinear regions, we also elaborated the Area Row and Column Structurer level to include explicit awareness of the relationship of size and number of area units as a direct variation, awareness of the capacity for converting commensurate area units, and the awareness of the coordination of operations for decomposing and recomposing partial units to make whole units. For example, during a Grade 4 teaching episode, we gave each child in the Midwest cohort drawings of two different blobs printed on a piece of paper with three commensurate grids on clear overhead projector sheets. The researcher asked each child to compare the area of the small blob to the area of the large blob. If the child responded that one was larger (or smaller) than the other, the researcher responded, "How much larger?" or "How much smaller?" We designed the task to motivate students to invent and use transitional units to link very small units to larger units, thereby allowing the comparison of vastly disparate regions in the same plane (e.g., how much larger is a given vast lake than a given small puddle?). We set up the following account of children's thinking by asking them to compare two very differently sized blobs.

Arielle, Abby, and Owen used the grids to count every square that was entirely or partially inside the blob (see Figure 6.2a for an example); Danny used the grids to count only squares that were completely inside the blob (see Figure 6.2b for an example); Anselm and Drew counted squares that were completely inside the blob but then decomposed and recomposed squares to account for partial units (see Figure 6.2c for an example).

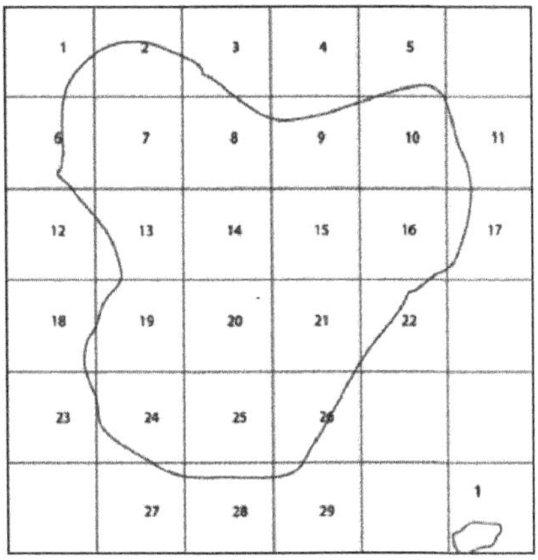

Figure 6.2a. Representation of counting squares that had any part inside the blob.

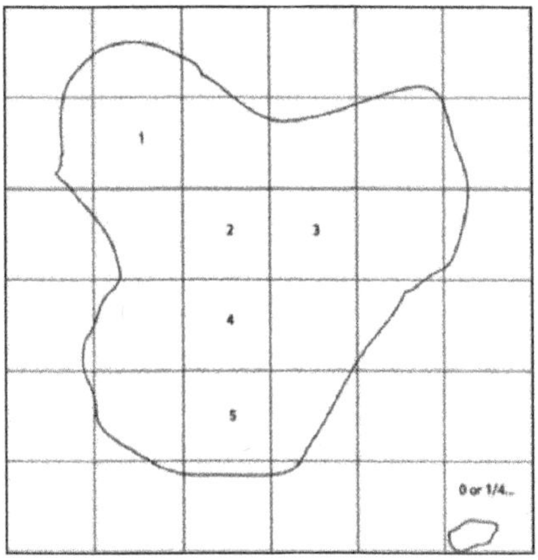

Figure 6.2b. Representation of counting squares completely inside the blob.

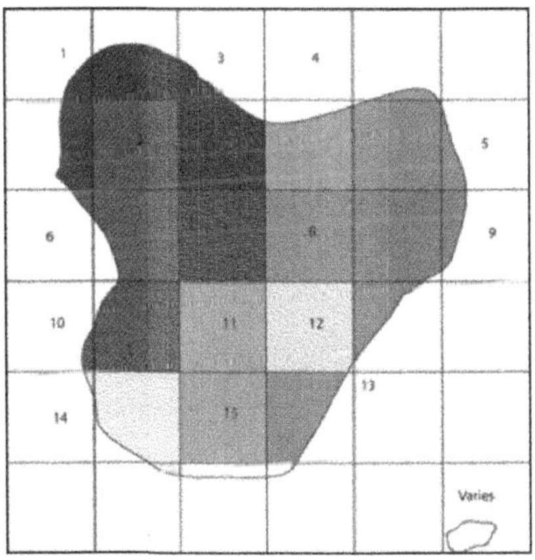

Figure 6.2c. Representation of decomposing and recomposing squares to account for partial units.

Although the children reported widely varying measures of area, all the children at the Midwest site were able to use the grids to relate commensurate area units and coordinate multiple units through multiplicative operations (for more information on curvilinear tasks and related findings, see Barrett et al., 2011; Cullen, Miller, Wikowski-Rumsey, Barrett, & Sarama, 2011). Because children at the Area Row and Column Structurer level used the dimensions to coordinate linear measurements and to relate linear and area units, we modified the description of the Area Row and Column Structurer level to include decomposing and recomposing partial squares to make whole units, as well as area unit conversions.

Deleting Area Conserver

The initial developmental progression of the LT for area measurement described children at the Area Conserver level as being able to reason about how regions that look different can have the same measure and to reason about area invariance. Findings in the Rasch modeling confirmed that reasoning about area invariance develops over time (see Figure 3.1 in Chapter 3); however, the results also indicated that the Area Conserver level spanned from below the Partial Row Structurer level and into the range of difficulty for the Array Structurer level. Although reasoning about area invariance is related to area measurement, insufficient evidence exists for us to assert that reasoning correctly about area invariance develops after Partial Row Structurer and before Area Row and Column Structurer. Qualitative evidence from the Midwest site supports this finding (see Area Conserver level in Chapter 5; Miller, Kara, Eames, Cullen, & Barrett, 2012). This evidence helped us decide to remove the Area Conserver level from the developmental progression of the LT for area measurement (see Table 6.1) and instead to address it as a vertically integrated component of several levels. We believe that children develop increasingly more sophisticated ways to reason about area invariance throughout the LT for area measurement, and so we have chosen to include children's understanding of area conservation across several levels, beginning at least as early as the Area Unit Relater and Repeater level and continuing on through the Area Row and Column Structurer level.

Table 6.1
A Comparison of the Levels From the Initial and the Revised Developmental Progressions for Area Measurement

Initial Developmental Progression		Revised Developmental Progression
Pre-Area Quantity Recognizer		
Area Simple Comparer	→	Area Quantity Recognizer
Side-to-Side Area Measurer	→	Physical Coverer and Counter
Primitive Coverer	→	Complete Coverer and Counter
Area Unit Relater and Repeater	→	Area Unit Relater and Repeater
Partial Row Structurer	→	Initial Composite Structurer
Row and Column Structurer	→	Area Row and Column Structurer
Area Conserver		
Array Structurer	→	Array Structurer
		Conceptual Area Measurer

Revisiting Array Structurer

By Grade 5, some children in the Midwest cohort showed evidence of thinking at the Array Structurer level, thereby confirming its existence and placement. Our results indicate that children operating at the Array Structurer level can imagine, describe, and use row and column structuring in addition to finding areas without making a drawing because their mental image is of a spatial or symbolic array. Children can compute area from the length and width of rectangles. At this level of thinking, children grasp that an array consists of a row of units taken as many times together as there are units along the adjacent side of that rectangle. That is, one must find two linear magnitudes to compute an area measure: The first describes the extent of one row of units, but the second is more complex. It describes the number of columns of square units that are filling up the entire array, with each unit along this rectangle edge representing an entire column of units in its place.

At times, some of the more advanced Grade 5 children recognized how the area of a rectangle related to the areas of other shapes, such as right triangles, trapezoids, and parallelograms. We took this understanding as initial evidence of a level beyond Array Structurer, which we investigated and then named *Conceptual Area Measurer* to indicate reasoning that supersedes the use of arrays.

Finally, we hypothesize the existence of thinking that is more advanced than the Conceptual Area Measurer level. Operations involving noncongruence transformations to measure areas under curves may characterize such thinking. These ideas are essential for advanced mathematics, such as integral calculus. More research is necessary to explore the development of these concepts beyond the previous levels of the LT.

Relating Area Measurement and Multiplicative Thinking

Throughout Chapters 3–6, we have restricted our focus to children's development of area measurement concepts. However, we should not study this domain in isolation from other geometric and mathematical aspects of development. We next discuss children's development in a related domain, multiplicative reasoning with whole numbers. Specifically, we describe a theoretical connection between children's development of multiplicative thinking and children's development of spatial reasoning for area measurement. This section summarizes our theoretical hypothesis regarding the interactions between the two domains. Area measurement differs from length measurement both mathematically and psychologically (Piaget, Inhelder, & Szeminska, 1960) because the area unit results from multiplying one-dimensional measures to create a two-dimensional measure (and similarly for volume). In this section, we discuss three aspects of multiplicative thinking in relation to the developmental progression of the LT for area measurement.

The first multiplicative aspect develops at the Area Unit Relater and Repeater level. Just before the Area Unit Relater and Repeater level, at the Complete Coverer and Counter level (formerly Primitive Coverer), children draw a complete covering, but their drawing and counting actions (e.g., drawing individual shapes one by one and then counting those individual shapes one by one) are not always systematic. The counting actions characteristic of the thinking at the Complete Coverer and Counter level, predominantly unsystematic drawing and counting the parts of a region being measured, do not appear to correlate with the foundational counting schemes requisite for multiplication as repeated addition. However, children at the Area Unit Relater and Repeater level have developed systematic drawing and counting actions (i.e., drawing individual units one by one and then counting those individual units systematically one by one) that correspond closely to the schemes of a simple multiplicative structure. That is, they have developed a type of multiplicative structure that is analogous to *n* times 1 square unit.

At the Initial Composite Structurer level (previously Partial Row Structurer), children have developed the capacity to build and manipulate a composite unit. They can structure a row as a composite unit consisting of individual units that repeat, but they do not necessarily structure the region in a comprehensive way. This type of multiplicative structure is analogous to *n* rows times *m* units per row. Because this application of a composite unit is not yet comprehensive, the children may not have generalized the symmetric operation of *m* columns of *n* units per column to the extent that they can set the first linear measure as the rows and then arrange columns, reasoning by an arbitrary assignment of one linear dimension as the rows and the other as the columns.

Finally, by the time that children reach the Array Structurer level, they can compute area from the length and width of rectangles, effectively imposing a symbolic array onto regions. This abstraction supports a fully multiplicative structure. That is, a child at the Array Structurer level can mentally structure an *n*-by-*m* array with the additional capacity to decompose such an array into constituent rows or columns to justify the measure of area as *nm* square units. Researchers need to conduct more work to verify these hypotheses about the interactions among multiplicative structures and the levels of the LT for area measurement.

REVISED DEVELOPMENTAL PROGRESSION

Figure 6.3 shows the revised developmental progression of the LT for area measurement. We characterize children's thinking in three parts and use three columns in the figure. In the first column, we present the levels of the developmental progression and provide prototypical examples of behaviors for that level. In the second column, we describe the mental actions on objects, including schemes that support the development of area measurement, as well as psychological foundations for quantifying area. In the third column, we describe the misconceptions or partial conceptions that children may exhibit at that level.

Developmental Progression	Mental Actions on Objects	Anticipated Misconceptions or Partial Conceptions
Area Quantity Recognizer (AQR) Identifies area as an attribute. *Comparing.* Uses side-matching strategies in comparing areas.	Perceives two-dimensional space and objects within the space. Using perceptual objects, employs internal innate competences to compare extent of regions. Because of the salience of linear extent, requests to compare regions may instantiate a process of comparing linear dimensions (Beilin, 1982) or intuitive and implicit summations of the linear dimensions of the objects (Bausano & Jeffrey, 1975; Maratsos, 1973; Mullet & Paques, 1991; Piaget et al., 1960; Ravn & Gelman, 1984; Russell, 1975; Sena & Smith, 1990).	*Tiling.* When covering a region with physical tiles, may leave large gaps between tiles or allow tiles to overlap one another or the boundary. *Drawing.* May unsystematically draw approximations of circles or other figures but attend to the constraints of the bounded region. 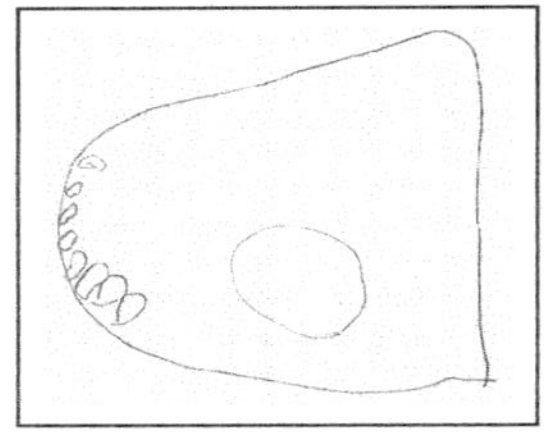 *Comparing.* May employ side-matching strategies in comparing areas by mentally or physically aligning the two objects and using only one dimension to compare. Given two rectangles, the child may intuitively identify that a larger region will require more squares to cover but may not accurately quantify the amount. May not recognize two rectangles as having equal areas when oriented or constructed differently. *Producing.* Builds (with gaps and overlaps) with foam squares, associating more squares with terms like *big* and fewer squares with terms like *small*.

Figure 6.3 continues on next page

Developmental Progression	Mental Actions on Objects	Anticipated Misconceptions or Partial Conceptions
Physical Coverer and Counter (PCC) Attends to some aspects of the structure. *Tiling.* Completely covers a rectangular space with physical tiles. *Comparing.* Makes intuitive comparisons of two-dimensional regions with simple direct comparisons. *Places one sheet of paper over another piece of paper to select the sheet that covers more space.*	With perceptual support, can visualize that regions can cover other regions. With strong guidance from prestructured materials, can direct the covering of that space and recognize that covering as complete. Can represent figural units as unanchored approximately rectangular shapes, aligning them (applied concept of collinearity), but often only intuitively and in one dimension (Mullet & Paques, 1991) or using height + width rules (Cuneo, 1980; Rulence-Paques & Mullet, 1998).	May not organize, coordinate, and structure two-dimensional space without perceptual support. *Quantifying.* When counting to a total, often relies on existing guides to direct counting but may count unsystematically in the interior. *Drawing.* When representing a rectangular tiling task, may initially draw approximately rectangular shapes, often leaving overlaps and gaps and often aligning only in one dimension. *Comparing.* May use superposition in addition to side-matching strategies when comparing two objects. 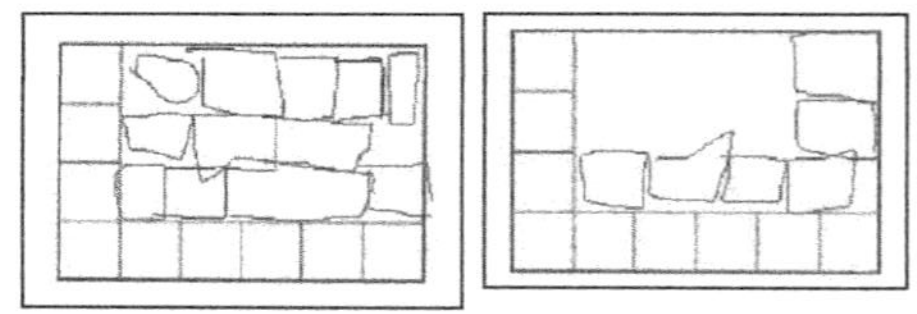

Figure 6.3 continues on next page

Developmental Progression	Mental Actions on Objects	Anticipated Misconceptions or Partial Conceptions
Complete Coverer and Counter (CCC) *Drawing.* Draws a complete covering of a specific region without gaps or overlaps and in approximations of rows. *Producing.* When given more than the total number of physical tiles needed, builds a region of specified area. *Builds a rectangle with an area of 12 from a pile of 20 tiles.*	Applies explicit understanding that must cover entire region with approximately rectangular shapes (in most rectilinear contexts). Implicit visual patterning of multiple concatenations of rectangles (and constraints of physical materials in physical tiling tasks) guides placement of anchored units in rows, approximately aligned with parallel rows. Employs counting-all action schemes (without using rows and columns) to enumerate the covering of a region. *Notes on psychological foundations: Not yet able to understand the basis for the multiplicative transformation of lengths into area measurements.*	*Quantifying.* In counting to a total, may count around the border and then systematically on the inside (possibly in a spiral, in an S-pattern, or in approximations of rows). Initially, may not accurately count shapes, losing track or double counting. Later, may not lose track and may count more accurately, but without row or column organization. 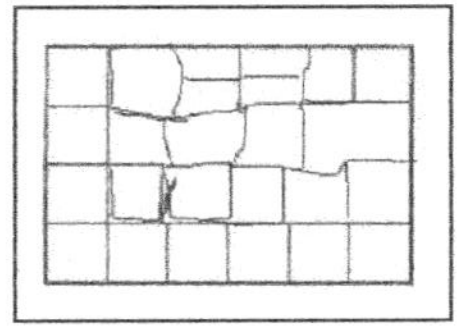 *Drawing.* When drawing approximations of rows, often has errors in alignment of the shapes and does not yet recognize the need for units of equal size. *Tiling.* When covering a region with physical tiles, may cover the whole region but use tiles of varying dimensions. *Iterating.* When covering a region without a complete set of physical tiles, may attempt to iterate but is unable to maintain equal iterations.

Figure 6.3 continues on next page

Developmental Progression	Mental Actions on Objects	Anticipated Misconceptions or Partial Conceptions
Area Unit Relater and Repeater (AURR) *Quantifying.* Guided by rows, counts individual units to measure. *Drawing.* Draws a complete covering based on an intuitive row or column structure. Children attend to drawing equal-sized units, one at a time. *Comparing.* Relates size and number of units (few larger, more smaller). Recognizes that different area units will result in different measures. Also recognizes the need for using identical units, at least intuitively or in some situations. Uses simple doubling strategies. Compares areas by counting units. *When researcher asks whether Shape A and Shape B take up the same amount of space on the paper, a child states that they take up the same amount of space "because they both have 4."* 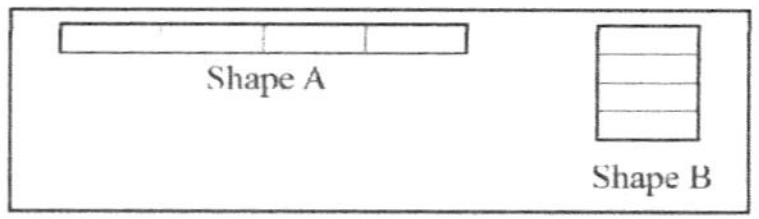 *Iterating.* When researcher gives fewer than the total number of physical tiles needed, child iterates individual tiles to measure and eventually cover a given region. *Producing.* Builds a region of area from an insufficient number of unit tiles through individual unit iteration (e.g., "leapfrogging"; when given a set of n tiles, the child may translate the first tile to represent both Tile 1 and Tile $n + 1$, the second tile to represent both Tile 2 and Tile $n + 2$ . . .).	Stronger constraints on object counting (e.g., counts all objects once and only once), and use of rows as an intuitive structure or explicit application of labeling as marker facilitates accurate keeping track. *Notes on psychological foundations: This level is foundational to the subsequent development of the multiplicative transformation of lengths into area measurement. This type of multiplicative structure is analogous to n times 1 square unit (e.g., to think of 12 square units drawn one at a time and counted one at a time as 12 times 1 square unit). At this level, educators would not expect a child to notice and represent collections of rows along a column to find an array.*	*Drawing.* When drawing, complete coverings may have some errors of alignment.

Figure 6.3 continues on next page

Developmental Progression	Mental Actions on Objects	Anticipated Misconceptions or Partial Conceptions
Initial Composite Structurer (ICS) Identifies a square unit as both a unit and a component of a unit of units (a row, column, or group); however, the child needs figural support to structure the space (possibly including physical motions of some of the tiles or drawing some collections of units rather than from using the dimensions). Reasonably estimates areas of regions. May apply an upper-bound (counting all wholes and partials as wholes) or lower-bound (counting only wholes as wholes) strategy. Tacitly recognizes usefulness of dimension displays as an indicator of the number of units along one dimension (row or column). *When drawing a rectangle of specified area, may succeed in identifying dimensions of a region without correctly drawing the array of units.* Organizes counting, drawing, or moving objects in composite units (units of units). • Sublevel A. Early on, may be unaware of the congruence of rows, yet uses a unit of units (not necessarily a column or a row) to determine a measure of a rectangular region. Structuring is very limited in scope; uses a mixture of individual units and units of units. • Sublevel B. Later, structures a rectangular region as a set of rows (more comprehensively than before), attending to the collinearity of rows (expects rows to have the same number of units). Does not yet coordinate rows and columns of units with the linear extent of both length and width along the measured object.	Builds, maintains, and manipulates mental images of composite units (e.g., conceptualizing and anticipating their existence or construction before they are perceptually present), structuring them as composites of individual shapes as a single entity—a row (a unit of units recognized as such). Applies this composite unit repeatedly (built and repeated at least once), but not necessarily exhaustively because intuition guides its application. *Notes on psychological foundations: This level anticipates later development of the multiplicative transformation of lengths into area measurement. This type of multiplicative structure can be analogous to n rows times m square units per row (e.g., for 12 square units drawn or counted, in part, as 3 rows with 4 square units per row).*	Organizes counting, drawing, or moving objects in composite units (unit of units), yet without a consistent application of that structure. Early on, may be unaware of the congruence of rows, yet uses a variety of rows to determine a measure of area. Later, structures the rectangle as a set of rows, understanding the collinearity of rows and the constraint that each row must have the same number of units. *Drawing.* May draw several rows by using parallel line segments but reverts to drawing individual squares. May begin drawing by repeating individual square units and curtail this process to draw parallel line segments to demarcate rows. *Comparing.* When comparing noncongruent regions with equal areas, may not recognize that they have the same areas. Confuses features of a region and measures of that region. For example, when asked whether Shape A and Shape B take up the same amount of space on the paper, states "They are the same because they both have four, but Shape A covers more space on the page." 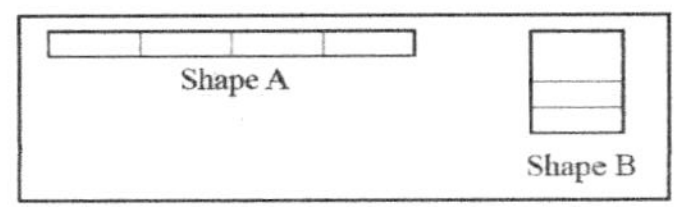 Initially, may ignore number labels or other dimension demarcation displays while drawing collections of units. Later, may coordinate only one dimension, the width or the height of the unit and the region.

Figure 6.3 continues on next page

Developmental Progression	Mental Actions on Objects	Anticipated Misconceptions or Partial Conceptions
Area Row and Column Structurer (ARCS) Decomposes and recomposes partial units to make whole units. *Drawing.* Uses the dimensions to constrain the placement of parallel row and column line segments. Measures both dimensions to determine the size of the iterated squares or rows of squares, as well as to determine the number of rows needed in drawing. May not need to complete the drawing to determine the area by counting systematically while attending to rows (most younger children) or computation (repeated addition or multiplication). *Comparing.* Explicitly relates size and number of area units (conversion). *Producing.* When drawing a rectangle of specified area, connects the measure of area with a count of area units without connecting the count of area units to the dimensions of the region.	Builds, maintains, and manipulates mental images, as supported by drawings, of composite units; has moved from identifying squares as individual units to seeing groups of squares as a component of a unit of units (e.g., a row and a column). Applies this composite unit repeatedly and exhaustively to fill the array (or instantiating this operation symbolically), coordinating this movement in 1:1 correspondence with the elements of the orthogonal column. In a measurement context, applies the concept that the length of a line specifies the number of unit lengths that will fit along that line, but the child may need to create a perceptual array to support such reasoning. This scheme may not be reversible. May connect the meaning of area measurement to a count of area units without relating those area units to linear units. *Notes on psychological foundations: This level is foundational to the later development of the multiplicative transformation of lengths into area measurement.*	*Drawing.* May succeed in identifying dimensions but needs to draw the array of square units to check. For example, may realize that a rectangle having an area of 16 square inches has 16 squares in it but struggles to determine the linear dimensions of that rectangle. *Quantifying.* May form an overgeneralization of the rectangular area formula, such as referring to this operation as *n* rows times *m* columns.

Figure 6.3 continues on next page

Developmental Progression	Mental Actions on Objects	Anticipated Misconceptions or Partial Conceptions
Array Structurer (AS) Has an abstract understanding of the rectangular area formula. *Quantifying.* Solves for area without making a drawing. In multiple contexts, child computes the area from the length and width of rectangles and explains how that multiplication creates a measure of area. The child operates arithmetically on area measures (e.g., comparing a 2 × 3 rectangular unit to an 8 × 10 rectangular region by finding the area of both and dividing the larger by the smaller for a ratio). *Comparing.* When comparing noncongruent regions with equal areas, realizes that the regions have the same area. Uses transitive reasoning.	Builds, maintains, and manipulates mental images of composites (units of units of units), such as a row of squares or rectangles composed of triangular units. Mentally decomposes and recomposes array into rows by columns or individual squares. The mental image may be of a spatial array or, at this level especially, a symbolic array. Area formula is reversible and decomposable: able to decompose a measure of area into a product of lengths and understand that those lengths indicate the number of area units in a row. Has an internalized mental image of the array. Can view each square as a unit, a component of at least two units of units (a row and a column) and the unit of units of units—the array. Integrates subdividing, covering, and numerical schemes; mental actions on objects match figural and physical actions on objects. *Notes on psychological foundations: The previous levels were foundational to the development of multiplicative transformation of lengths into area measurement, which develops at this level. This type of multiplicative structure is analogous to n rows times m units per row and m columns times n units per column. Note that a more global spatial-structuring scheme would enact the symmetrical operation.*	*Drawing.* When working within nonrectangular regions, may struggle to solve for area without making a drawing. *Quantifying.* May not recognize that formulas for areas of these shapes relate to the formula for the area of a rectangle.

Figure 6.3 continues on next page

Developmental Progression	Mental Actions on Objects	Anticipated Misconceptions or Partial Conceptions
Conceptual Area Measurer (CAM)		
Has an abstract and generalizable understanding of the rectangular area formula.	Bases numerical operations on measures on abstracted spatial structuring.	
Quantifying. Restructures regions to determine how to use known area measures to find the areas of triangles, kites, trapezoids, and parallelograms. Recognizes that formulas for areas of these shapes relate to the formula for the area of a rectangle. Provides arguments and justifications as evidence of physical or figural decomposition or recomposition and rigid transformations of these shapes. Uses geometric properties of these shapes to support reasoning.	Can coordinate and operate on geometric properties of shapes to develop formulas for areas of triangles, kites, trapezoids, and parallelograms. The child physically or figurally decomposes, recomposes, and transforms these geometric shapes to relate formulas.	
Comparing. When comparing noncongruent regions with equal areas, integrates and operates on qualitative and quantitative aspects of the regions.		

Figure 6.3. Revised developmental progression component of the LT for area measurement.

SECTION III

Chapter 7

Volume

Douglas H. Clements
University of Denver
Julie Sarama
University of Denver
Douglas W. Van Dine
Metropolitan State University of Denver

Volume presents a significant challenge to children's geometric and spatial structuring competences because it requires children to coordinate measurements involving three dimensions. Further, volume adds complexity because of the nature of the materials used in measuring it. That is, we often pack with solid units, such as when we view the volume of a rectangular prism as measured by a three-dimensional array of cubes. In comparison, a liquid may fill three-dimensional space, taking the shape of the container. In this chapter, we synthesize some of the related research and then present the developmental progression of the initial learning trajectory (LT) for volume measurement.

We define volume as the amount of three-dimensional space that a boundary contains. Two assumptions of volume measurement are that a three-dimensional region (e.g., a cube) is the unit and that the volume of the union of two nonoverlapping or disjoint spaces is the sum of their volumes. In contrast to this abstract and general definition, however, children's experiences of volume differ from the aforementioned packing-versus-filling situations. Curry and Outhred (2005) concluded that for children in Grades 1 through 4, a similarity exists between competence in filling volume and competence in length measurement. This finding is consistent with the notion that children often view filling tasks as one-dimensional. For example, children may believe that the height of the liquid changes rather than that the volume changes. As a result children may think about volume in terms of the height of the liquid in the container (i.e., a length) and view filling as iterating the height of that liquid, either mentally or physically, up the side of the container. This perception is further evidence that filling volume may be phenomenologically one-dimensional for young children. That is, children initially learn to make meaning by using one-dimensional strategies. Nevertheless, researchers have also found that some children performed better on length than on filling volume, whereas others performed better on filling volume (Curry & Outhred, 2005). This result raises the question of whether a relationship exists between the learning of length and the learning of filling volume. For example, they may be mutually reinforcing, but they appear to have distinctive cognitive foundations in some cases; previous learning of length could also conceivably hamper the learning of filling qua volume.

Distinct from filling volume, which can be reduced to a single dimension, packing volume appears to demand attention to all three dimensions. Curry and Outhred (2005) conjectured that children find packing volume tasks not only more difficult than length and area tasks but also more difficult than filling volume tasks. Although their research found substantial increases in children's competence with packing tasks from Grade 2 to Grade 4, the older children's performance tended to remain below the corresponding performance for area tasks. These results lend support to the claim that an understanding of spatial structuring of area is prerequisite to understanding packing volume.

The measurement of packing volume requires that children build another component onto their competence for spatial structuring. Work on spatial structuring describes a variety of strategies that children use to structure three-dimensional arrays of cubes (Battista & Clements, 1998). Younger children frequently count only the faces of the cubes, a process that often results in counting some cubes (e.g., those at the corners) more than once and not counting interior cubes. Battista and Clements found that most fifth graders, but only 20% of third graders, structured the cubes constituting a rectangular prism as a series of layers, each of which is structured into rows and columns (similar to the two-dimensional structuring for area previously discussed). This spatial structuring strategy allows children to determine the number of cubes in one layer and then to multiply or skip-count to obtain the total number of cubic units in the rectangular prism.

As in the development of an understanding of length measurement, the inverse relationship between unit size and reported measure in volume is an important concept for children. Sophian (2002) examined young children's understanding of volume and reported that preschoolers could, with instruction, reason that they would need fewer large objects than small objects to fill a container. She found that children improved significantly from a pretest to a posttest when they had experience filling containers by using objects of different sizes.

This limited body of research formed the basis for the initial developmental progression of the LT for volume measurement (Sarama & Clements, 2009b). For example, we identified separate levels for filling volume (i.e., capacity) and packing volume (i.e., three-dimensional arrays), placing capacity before packing volume; and we extrapolated levels from our developmental progression for length (e.g., Length Direct Comparer, Length Unit Relater and Repeater) because of extant research that suggests a structural isomorphism between length and capacity (Curry, Mitchelmore, & Outhred, 2006; Curry & Outhred, 2005). We therefore hypothesized that the mental actions on objects for filling volume were similar to those for length; the development would also be structurally similar and would likely reflect previous learning of length. That is, the developmental progressions within one domain (i.e., length) repeat themselves in this new linear context (see an explanation for this phenomenon as a principle of the theory of hierarchic development in Sarama & Clements, 2009b, p. 21).

For packing volume, previous work on spatial structuring dominated the initial developmental progression. As with area, we hypothesized a structuring process in which children mentally construct the cube as a unit of volume, then units of units (i.e., rows of cubes), iterated and forming units of units of units (i.e., layers of cubes), which, iterated, constitute the volume of a rectangular prism. We also hypothesized that these constructions are essential to children's conceptual understanding of volume measurement, as well as the meaningful use of formulas. In this way, the developmental progression also embodied the notion of mutual development of concepts and skills.

Figure 7.1 presents the developmental progression of this initial LT, along with the mental actions on objects that we hypothesized to underlie the behaviors indicative of each level of thinking (Sarama & Clements, 2009b). This initial LT was the starting point for our research; however, from the preliminary research review (Clements & Sarama, 2007; Sarama & Clements, 2009b) to the beginning of the longitudinal study reported in this monograph, we were as aware of the limitations of the research base for volume as we were for area. Again, a limited number of studies of children's learning of volume existed, and the notion of spatial structuring tended to dominate those studies (e.g., Battista & Clements, 1996). As with area, we believe that spatial structuring is a critical component of understanding volume. However, we also believe that other components have not received enough attention from researchers. In the final section of this chapter (the quantitative analyses) and in the three chapters that follow, we describe our efforts to test, refine, and extend the developmental progression of the initial developmental progression of the LT for volume measurement.

Developmental Progression	Actions on Objects
Volume Quantity Recognizer Identifies capacity or volume as attribute. *Says "This box holds a lot of blocks."*	Perceives space and objects within the space.
Capacity Direct Comparer Can compare two containers. *Pours one container into another to see which one holds more.*	Using perceptual objects, internal bootstrap competencies compare linear extent or recognize overflow as indicating that the container from which something is being poured contains more than the container into which something is being poured.

Figure 7.1 continues on next page

Developmental Progression	Actions on Objects
Capacity Indirect Comparer Can compare two containers by using a third container and transitive reasoning. *Pours one container into two others one at a time and concludes that one holds less because it overflows, whereas the other is not entirely filled.*	The child has a mental image of a particular amount of material that can be built, maintained, and manipulated. With the immediate perceptual support of the containers and material, the child can compare such images. Some may apply explicit transitive reasoning to the images or to their symbolic representations.
Primitive 3-D Array Counter Partial understanding of cubes as filling a space. *Initially, may count the faces of a cube building, possibly double counting cubes at the corners and usually not counting internal cubes.* *Eventually counts one cube at a time in carefully structured and guided contexts, such as packing a small box with cubes.*	With perceptual support, can visualize that filling 3-D space with objects (e.g., cubes) is possible. With strong guidance and perceptual support from prestructured materials, can direct the filling of that space and recognize that filling as complete, but often only intuitively. Implicit visual patterning and constraints of physical materials guide placement of cubes.
Capacity Relater and Repeater Uses simple units to fill containers, with accurate counting. *Fills a container by repeatedly filling a unit and counting how many.*	See the Length Unit Relater and Repeater LT level.

Figure 7.1 continues on next page

Developmental Progression	Actions on Objects
Partial 3-D Structurer Understands cubes as filling a space, but does not use layers or multiplicative thinking. Moves to more accurate counting strategies. *Counts unsystematically, but attempts to account for internal cubes.* *Counts the numbers of cubes in one row or column of a 3-D structure and uses skip counting to find the total.*	Builds, maintains, and manipulates mental images of composite shapes, structuring them as composites of individual shapes and as a single entity—a row (a unit of units), then a layer (a unit of units of units). Applies this composite unit repeatedly, but not necessarily exhaustively, because intuition continues to guide its application.
3-D Row and Column Structurer *Counts or computes (row by column) the number of cubes in one row and then uses addition or skip counting to determine the total.* *Computes (row times column) the number of cubes in one row and then multiplies by the number of layers to determine the total.*	Builds, maintains, and manipulates mental images of composite shapes, structuring them as composites of individual shapes and as a single entity—a layer of congruent squares. Applies this composite unit repeatedly and exhaustively to fill the 3-D array—coordinating this movement in 1:1 correspondence with the elements of the orthogonal column. If in a measurement context, applies the concept that the length of a line specifies the number of unit lengths that will fit along that line.
3-D Array Structurer With linear measures or other similar indications of the two dimensions, multiplicatively iterates squares in a row or column to determine the area. *Constructions and drawings are not necessary. Computes the volume of rectangular prisms from their dimensions and explains how that multiplication creates a measure of volume.*	Builds, maintains, and manipulates composites (a 3-D array—a unit of units of units of units) that operate in two dimensions. Mentally decomposes and composes a 3-D array into layers, which themselves are decomposed or composed into rows by columns. The mental image may be of a spatial array or at this level especially, a symbolic array. Applies repeated addition or multiplication to composites.

Figure 7.1. Initial developmental progression of the LT for volume measurement (adapted from Sarama & Clements, 2009b).

As presented in the previous chapters, we used Rasch modeling of our final assessment data ($n = 258$) to provide quantitative evidence for verifying and refining the developmental progression from the initial LT. The Rasch model indicated unidimensionality of the volume construct, with the measures explaining 61.9% of the raw variance, an eigenvalue of 1.6, and 4.9% of unexplained variance in the first contrast (acceptable values are above 40%, under 2.0, and 5%, respectively). Evidence of volume measurement as a unidimensional construct fulfills one of the underlying assumptions of Rasch modeling, indicating that Rasch modeling is appropriate for this data. The range of infit MNSQ for items was between 0.74 and 1.32, within the range of productive measurement, 0.5 to 1.5 (Linacre, 2014). Thus, all items on the instrument align well, and we can assume that they measure the same construct.

As mentioned in Chapter 1, item reliability is an estimate of the replicability of item placement within the ordered list of items along the scale and is conceptually similar to Cronbach's alpha. Item separation indices are estimates of the distance between items. The separation index for the Rasch modeling of the volume items suggested that the item difficulty hierarchy was suitable for the sample used in this study, and the assessment was able to distinguish between low-performing children and high-performing children. As with length and area, a low person separation index of 1.04 and a reliability of 0.52 were inconclusive for volume and suggested a low likelihood of replicability in children's samples outside the final assessment sample. The item–person map, however, further confirmed that the spread of item difficulty alongside children's ability was consistent with the LT for volume measurement.

Our purpose in this section was to verify and provide evidence for refining the developmental progression of the LT for volume measurement by using quantitative data. Figure 7.2 depicts all items arranged with item-difficulty estimates on the vertical axis and corresponding level intervals marked to represent the range of difficulty for each pair of items designed to assess the same level of understanding, following the same process used for length and area. Again, each interval, moving from left to right in the figure, corresponds to an increasingly sophisticated level, as hypothesized in the developmental progression of the initial LT. In general, the pattern of these intervals suggests a progression that is consistent with our hypothesized developmental progression. That is, an increase in item difficulty corresponds to a move to more sophisticated levels in the developmental progression of the LT with one exception, that of Capacity Relater and Repeater.

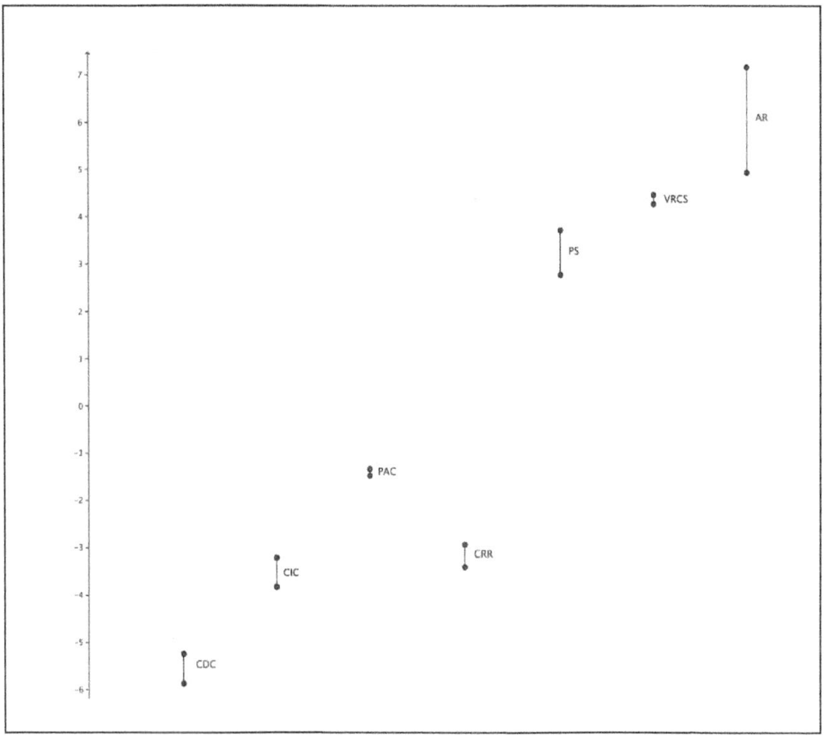

Figure 7.2. Item difficulty aligned with the levels of the LT for volume measurement. CDC = Capacity Direct Comparer, CIC = Capacity Indirect Comparer, PAC = Partial 3-D Structurer, CRR = Capacity Relater and Repeater, VRCS = 3-D Row and Column Structurer, AR = 3-D Array Structurer.

These results provide support for the developmental progression from the initial LT for volume measurement overall. Nonetheless, further revision is necessary in the assessment of items and in our theory to increase reliability and to refine, reposition, or revoke the Capacity Relater and Repeater level. We return to these issues in Chapter 10.

Chapter 8

Volume in the Early Childhood and Primary Grades

Douglas W. Van Dine
Metropolitan State University of Denver
Douglas H. Clements
University of Denver
Julie Sarama
University of Denver
Jennifer S. McDonel
Radford University
Lisa Napora
Daemen College

This chapter presents qualitative analyses of young children's thinking and learning of volume. Throughout our 4-year longitudinal study, children in pre-K through Grade 2 exhibited actions on objects that were mainly consistent with the developmental progression within the learning trajectory (LT) for volume measurement (Sarama & Clements, 2009b). We present evidence from three children at the Northeast site (Edith, Robert, and Lia) for whom we have 4 years of longitudinal data. We also present complementary evidence from other children in the Northeast cohort as needed to fill gaps for various levels of the developmental progression. In addition, we highlight analyses that suggested that the initial developmental progression of the LT for volume measurement needed revision; however, we have reserved detailed discussion of the revisions for Chapter 10.

CONSISTENCY WITH INITIAL LEARNING TRAJECTORY FOR VOLUME MEASUREMENT: QUALITATIVE EVIDENCE

Analysis of the initial assessment indicated that pre-K children in the Northeast cohort already identified volume as an attribute. However, the nature of the clinical interview method used on the initial assessment did not afford an opportunity to probe children's thinking or verbalization competencies in depth. On the initial assessment, we also observed children using direct comparison strategies to compare capacities of containers and volumes of objects. During the longitudinal study, the children's spatial structuring skills increased while they progressed through various levels of volume quantification, including recognition of (a) cube faces; (b) cubes as blocks; (c) $1 \times 1 \times 1$ cubes as units; (d) $1 \times 1 \times n$ cores, rows, or columns (i.e., units of units); and then (e) $1 \times n \times m$ layers or slices (i.e., units of units of units). Furthermore, throughout the study, we observed the development of children's additive thinking and movement toward multiplicative thinking while structuring volume.

Volume Quantity Recognizer

Children at the Volume Quantity Recognizer level demonstrate an initial recognition of volume by identifying capacity or volume as an attribute and by perceiving space and objects within three-dimensional space. Children at this level make such statements as "That box holds a lot of blocks" and often use general terms such as *big* and *small,* but they are not necessarily able to relate their observations to the dimensions of the box or to quantify a number of blocks.

At the beginning of the study, in the spring of pre-K, both Edith and Robert demonstrated reasoning above the Volume Quantity Recognizer level. Lia and Marina, however, demonstrated thinking consistent with the Volume Quantity Recognizer level. For example, on one initial assessment task, we gave the children a small cube and asked them to predict how many cubes they would need to measure the space inside a juice box (Figure 8.1 shows materials for this task). Lia looked at the box and said, "You make some room and you put it in there," while she placed the cube inside the box, but she did not report a quantity or seek more cubes to fill the box. Marina picked up the box and looked inside, then announced, "2," indicating that she was considering the interior of the box but not relating the size of the cube to the volume of the box. We interpreted these behaviors as evidence that Lia and Marina perceived the space inside the box (i.e., that the box could hold objects) and that the cubes would take up space inside the box—both of which reflected mental actions on objects consistent with the Volume Quantity Recognizer level. Neither child displayed thinking more sophisticated than Volume Quantity Recognizer because their focus was simply on recognizing the interior of the box; neither Lia nor Marina sought to fill the box completely or to structure the cubes inside the box.

Figure 8.1. Box and cube used in an initial assessment task.

On the initial assessment, several children demonstrated thinking consistent with the Volume Quantity Recognizer level when we showed four objects (see Figure 8.2), told the children to imagine that the objects were made of ice cubes, and asked them to determine which object would melt into the largest puddle. Rhonda looked at the blocks, pointed to the 3 × 2 × 1 array and then the S-shape, and said, "This is small; this is big." On the same task, Lia ignored the L-shape, but then pointed to the other three solids from left to right, saying, "This one's fat, this one is large, and this one is tiny." We interpreted this thinking as indicative of the Volume Quantity Recognizer level because children made gross comparisons of the blocks and identified their volumes as an attribute. The children did not, however, attempt to align the blocks to compare them directly, discuss them in terms of any of their dimensions, or otherwise quantify the blocks.

Figure 8.2. Initial assessment "ice block" task.

Similarly, we observed children making gross comparisons of capacity. For example, in the fall of kindergarten, we presented children with cups of varying sizes and asked them which cup they would want if all cups were full of lemonade. On this task, Edith, Robert, and Lia each chose the cup with the largest capacity. When we asked Lia why she chose the blue cup (see Figure 8.3), Lia put her hand inside the top of the blue cup, then continued to the yellow cup, then to the green cup, then to the red cup, saying, "Cause that one is bigger, and this one is tiny, and this one is tiny, and this one is tiny." The only differentiation that Lia made was between the largest cup and the others, with no reference to dimensions. Like with the developmental progression for length, we interpreted these intuitive comparisons of capacity as further indication of the Volume Quantity Recognizer level, in which children made gross comparisons of length without physically aligning objects or verbalizing a mental alignment.

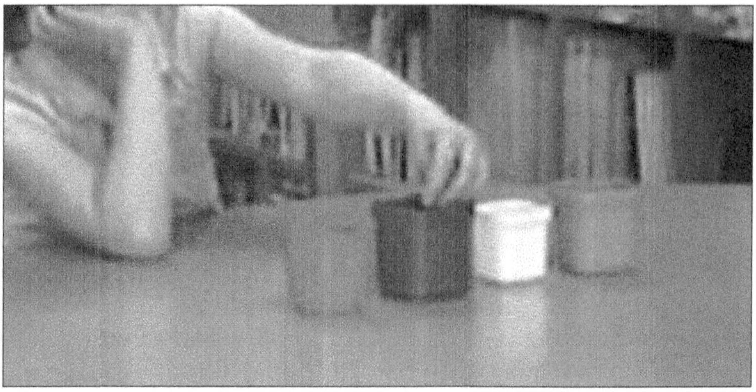

Figure 8.3. Lia chooses the largest cup for lemonade. (Note: From left to right in the photograph, cups are red, blue, yellow and green.)

These examples provided evidence that the children had an awareness of volume (or capacity) as an attribute at the beginning of the study; however, most of the children were beyond the Volume Quantity Recognizer level. The data also indicated that children treated tasks focused on filling differently from those focused on packing. Although much previous research focused on the initial phases of volume understanding as developing around the idea of capacity, we began to notice the need to expand the developmental progression to include not only filling volume but also packing volume even at these initial levels.

Capacity Direct Comparer

Children at the Capacity Direct Comparer level, who have progressed beyond simply recognizing volume as an attribute, demonstrate the ability to compare the volumes of two containers or objects. Children exhibiting thinking at the Capacity Direct Comparer level are able to compare the capacities of two containers, often through the action of pouring from one container into another. Initially, children may compare the containers by using perceptual cues and focusing only on linear extent (e.g., the height of cylinders; see Piaget & Inhelder's [1948/1967] work on conservation). Eventually, when pouring from one container into another, children can recognize overflow as indicating that the container from which they are pouring can contain more than the container into which they are pouring.

When we asked the children in our study to compare the volumes of objects and capacities of containers directly, the children physically manipulated objects, initially aligning them by the longest dimension and referring to linear extent alone when comparing, using such terms as *taller, bigger, shorter,* and *smaller* to describe differences—a shift away from gross comparisons but with no reference to dimensions, which are characteristic of the Volume Quantity Recognizer level. For example, as a follow-up to the initial assessment task in which children compared the volumes of the four solids (see Figure 8.2), we specifically asked children to compare the volumes of the L-shape and the $3 \times 2 \times 1$ array, each of which included six cubes. Robert laid the two solids so they were side by side on the table, put his hand on the L-shape, and said, "This one's larger." Similarly, Edith stood the two solids on the table next to each other to compare their heights and responded, "This one's larger" [holding up the L-shaped solid]; "This one's smaller" [holding up the $3 \times 2 \times 1$ array].

During the cups task shown in Figure 8.4, Marina was indecisive at first about which cup to choose. She chose the yellow cup and then changed her mind to the blue one.

Researcher: Why do you want the blue one?

Marina: When I do like this [pushing the cups together with her hands then laying her head down sideways to look across the tops of the cups], this one is short, medium [pointing to the yellow, then the green cups]. This one is small, medium, large [pointing to the yellow, green, then red cups] and two of these mediums.

Researcher: Two of those are mediums?

Marina: And this one is large [pointing to the blue cup].

Researcher: And why do you want the large one?

Marina: Because the large is the biggest.

Researcher: So does that mean that the blue one would hold the most lemonade?

Marina: Yes.

Although Marina struggled to find unique words to describe all the cups, we interpreted her actions as indicative of the Capacity Direct Comparer level because she physically aligned the cups, then visually compared and attempted to label them by one dimension (i.e., height), even though the other dimensions also varied.

Figure 8.4. Marina compares the cups by height, aligning them and looking across the tops. (Note: From left to right from the readers' perspective, cups are blue, red, green, and yellow.)

Also during the first teaching episode of the longitudinal study, some children began referring to more than one dimension in their capacity comparisons and often combined vocabulary with a motion or gesture to indicate the compared dimensions. For example, during a teaching episode in the fall of pre-K, we presented children with a graduated cylinder and a tall stemmed glass (see Figure 8.5) and then asked which one they would like if both were full of lemonade. Lia chose the glass,

> because that one is tiny [using her thumb and forefinger to refer to the opening at the top of the graduated cylinder]; this one is that big [referring again to the opening of the glass]. This one is bigger and that one is littler [sweeping her hand up the glass, then down the cylinder to indicate height].

Lia clearly referred to more than one dimension in her comparison, as indicated by her hand gestures. In doing so, she seemed to neglect her previous gross comparison of the two containers (i.e., that the glass had a stem for the bottom half that would not hold any liquid) in favor of comparing dimensions. Although Lia referred to two dimensions, we took this as evidence of the Capacity Direct Comparer level because her focus was on comparing volume by using only the linear extent.

Figure 8.5. Lia compares capacity by using more than one dimension.

Both Robert and Edith demonstrated thinking at the Capacity Direct Comparer level in tasks that involved pouring from one container into another during the fall of kindergarten. In one teaching episode, we presented the children with two graduated cylinders of different diameters and asked which one had the larger volume, after clarifying that *volume* meant how much each cylinder could hold. When we provided water, both Edith and Robert filled one cylinder, poured it into the other, and correctly determined the one with larger volume. In a second task, 2 weeks later, we presented the children with two different containers and rice as the filling material. Again, both correctly determined the container with the larger volume by filling one container and pouring it into the other.

In the fall of Grade 2, we again gave children tasks that asked them to choose which of two containers they thought had the larger volume without doing the direct comparison first, mainly as verification of a level they had shown previously. Children used different strategies for deciding which one was larger, and we encouraged them to make a choice. After a child had selected one container as "larger" (whether or not this determination was correct), we filled the "larger" container with water. We then asked the child what would happen if we poured the water from the filled container into the other one. Both Edith and Robert thought the "smaller" container would fill exactly to the top but would not overflow; Lia, however, said that she was unsure of what would happen.

Thus, although children could directly compare capacities at younger ages, they did not mentally "run" the procedure (cf. Simon, 2013) to anticipate the result during Grade 2, nor did we find evidence that they anticipated the outcomes of comparisons for capacity before Grade 2. Hierarchic interactionalism posits the continued existence of earlier levels, and the role of intentionality and social influences in their instantiation, which explains why in some contexts even adults fall back to earlier levels (e.g., fail to conserve in certain situations). The lack of experience with this problem context may have evoked earlier levels of thinking in this situation (see discussion of nongenetic levels in Chapter 1). Alternatively, the children did not reflect the physical actions of comparing capacities to a more abstract representation, perhaps because of a lack of articulation during the physical experiences. That is, their procedures for comparing capacities remained at the physical–concrete level because the children lacked experiences (i.e., tasks and interactions) that engendered the reflective abstraction necessary to rebuild the procedures at the integrated–concrete level that would support anticipation.

Our evidence supported the claim that children move into direct comparison shortly after recognizing volume as a quantity, so the Capacity Direct Comparer level is placed appropriately within the developmental progression of the LT for volume measurement. However, as with the Volume Quantity Recognizer level, we observed children's development not only in filling, but also in packing. We therefore see a need for expansion to include packing at this early level.

Capacity Indirect Comparer

An indirect comparison strategy is required in situations in which comparing two objects directly (that is, by physically or mentally aligning them) is not possible or convenient. For capacity, this strategy may involve pouring the contents of a third container into each of the first two containers and then using transitive reasoning to make sense of the results. From our observations, this thinking seemed to be a special strategy that children used on occasion and often only when researchers prompted them; most children used direct comparison whenever that option was available. In tasks that the researchers designed specifically to elicit behaviors indicative of indirect comparison, the children in the study almost always attempted to make visual direct comparisons instead; therefore, as discussed in Chapter 2, it was difficult to determine whether children were unable to compare indirectly or whether they simply chose not to. Consequently, the placement of indirect comparison within the developmental progression of the LT for volume measurement currently is undetermined.

Nonetheless, we did observe children inconsistently using indirect comparisons during the study. In a teaching episode during the fall of kindergarten, we presented children with several tasks that asked them to compare the volumes (i.e., capacities) of pairs of containers by filling them with rice or water. Specifically, we gave children a container filled with rice or water and asked whether they could use that container to compare the volumes or capacities of the other two; that is, we were asking if they could determine which of the two containers had the largest volume. In some of the tasks, the initially filled container had a volume or capacity that was smaller than that of either of the two containers that the children were comparing; and in other tasks, the initially filled container had a volume or capacity between those of the other two. During this teaching episode, both Edith and Robert consistently demonstrated transitive reasoning and were able to correctly compare the volumes indirectly. Lia, however, was not able to use the initially filled container to compare the other two. In addition, when the researcher took the lead in pouring from the filled container into the others, Lia did not demonstrate the ability to reason transitively about the volumes or capacities of the two containers that we asked her to compare.

Later, during a teaching episode in the spring of kindergarten, we presented Robert with a similar task, but this time we began by pouring one scoop of water into each container. Robert incorrectly identified the container in which the water reached higher after one scoop as the larger container. The researcher continued to put one scoop at a time into each container, asking Robert each time which container he thought was the larger one; Robert persisted in saying that the container in which the waterline was higher was larger. After the fifth scoop of water, the container that Robert indicated was larger overflowed with water although the other one was not yet full. Robert persisted in his determination that this container was larger than the one that was not yet full, even though he had consistently determined that the one that overflowed was the smaller container in previous teaching episodes.

Similarly, in a teaching episode 2 weeks later, Lia demonstrated an inability to compare capacities indirectly. When the researcher presented two containers, she visually inspected them and mentally determined that they were the same size. When the researcher gave her a scoop, she found that one container held five scoops of rice, whereas the other held six. Rather than change her determination that both were the same size, Lia attempted to describe the difference by indicating that one container was made of glass and the other of plastic. Although she demonstrated more sophisticated levels of volume understanding, Lia never consistently demonstrated an ability to compare capacities indirectly during the 4 years of the study.

Overall, Edith was the only child in the Northeast cohort who demonstrated thinking at the Capacity Indirect Comparer level throughout the study, and Lia was the only child who did not solidly demonstrate thinking at the Capacity Indirect Comparer level on any task. The other children in the Northeast cohort exhibited more variability in their thinking, as Robert did. That is, on some tasks, they demonstrated transitive reasoning and the ability to indirectly compare capacities; but on other tasks they used faulty reasoning, or they simply focused on comparing capacities directly through visual inspection. Further research is necessary to accurately determine the placement of indirect comparison of volume and capacity in the developmental progression of the LT for volume measurement. We posit that it may well develop in parallel with the other levels, depending on specific experiences.

Primitive 3-D Array Counter

Although the previous levels of Capacity Direct Comparer and Capacity Indirect Comparer describe children who are mainly producing qualitative comparisons, children at the Primitive 3-D Array Counter level begin to quantify the volume and capacity of three-dimensional objects. At the Primitive 3-D Array Counter level, children show a partial understanding of cubes as filling space. They are able to visualize that filling a three-dimensional space with cubes is possible; and with strong guidance and perceptual support, they can direct the filling and recognize when the filling is complete. When a researcher presents a child with an object structured from cubes and asks him or her to determine its volume, a child at the Primitive 3-D Array Counter level may initially count only the faces, possibly double counting at the vertices or edges; in addition, he or she does not account for hidden or internal cubes. With guidance and support, children eventually count one cube at a time in carefully structured contexts, such as packing a small box with cubes.

On the follow-up to the initial assessment, Zola demonstrated a transition between direct comparison and quantification strategies when we asked her to compare the volumes of the L-shape and the $3 \times 2 \times 1$ solid shown in Figure 8.2. She identified the L-shape as having a larger volume than the $3 \times 2 \times 1$ solid and then tried to count the number of cubes in each; but when her count did not match her visual comparison, Zola struggled to resolve the dissonance that she had noted.

Researcher: [Placing two sets of blocks on the table in front of Zola, one constructed of six cubes in an L-shape, and the other consisting of six cubes in a $3\times 2 \times 1$ rectangular prism] Let's look at these two [pointing alternately at each set of cubes]. Please compare these two sets of blocks for volume. Do they have the same volume, or is one larger?

Zola: This one [holding up L-shape] is larger than this one [holding up $3 \times 2 \times 1$ prism].

Researcher: How do you know that?

Zola: Because this one [holding up L-shape], it has more blocks than this one [holding up $3 \times 2 \times 1$ prism].

Researcher: This one [pointing at L-shape] has more blocks than this [pointing at $3 \times 2 \times 1$ prism]?

Zola: Yes.

Researcher: All right. Well, how do you know it has more blocks? Show me.

Zola: Because.

Researcher: How many blocks does it have?

Zola: [Pointing at and counting the cubes in the L-shape] Five. [Note: The L-shape actually includes six cubes.]

Researcher: Okay.

Zola: But this one [holding up $3 \times 2 \times 1$ prism] has more blocks than this one, but this is more taller [standing the L-shape up on the table]

	because…this one is short [standing the 3 × 2 × 1 prism on end], and this one is tall [standing the L-shape on end].
Researcher:	This one is short, and this one is tall, okay; but how many blocks does this one have? [pointing at L-shape]
Zola:	[Pointing and again incorrectly counting the cubes] Five.
Researcher:	Five. And how many blocks does this one have? [indicating the 3 × 2 × 1 prism]
Zola:	[Counts the six faces on one side then flips it over and counts the six faces on the other side] Twelve. This one has more blocks [holding up the 3 × 2 × 1 prism], but this one is the tallest [standing the two prisms side by side and placing a hand on top of each].

We interpreted this response to indicate that Zola had begun to recognize cubes as filling space and attempted to use the number of cubes as a means to compare the volumes. However, because she double-counted cubes and counted incorrectly, she was unable to resolve the disparity between her comparison by number and her comparison by height. Ultimately, she fell back to direct comparison by linear extent, a behavior consistent with the theory of hierarchic interactionalism. We recognize that children move into increasingly sophisticated levels of thinking without abandoning previous levels; rather, children may fall back to previous levels of thinking when convenient or necessary to address a complex task.

Robert demonstrated a foray into the Primitive 3-D Array Counter level on the initial assessment in pre-K during the first year of the study. On the initial assessment item depicted in Figure 8.1, Robert looked inside the small box and predicted that three cubes would fill it. When we gave him more than enough cubes, he placed cubes one at a time into the box, counting as he did so, and counted out four cubes while he placed them in the bottom layer, completely filling that layer. He did not, however, continue to place a second layer of cubes on top of that one to fill the box completely. Later in the same assessment, the researcher presented Robert with a 2 × 2 × 3 rectangular prism constructed of wooden cubes and asked how many cubes were in the prism. He pointed to and counted the individual squares in only the top face and said that six cubes were in the entire prism. We saw this response as evidence that Robert recognized the cubes as filling space but lacked a complete understanding of the space as consisting of a specific quantity of cubes. He counted faces rather than cubes, which is indicative of the Primitive 3-D Array Counter level.

On the same initial assessment task, Lia pointed at individual squares on the top face while she counted "1, 2, 3, 4, 5, 6, 7, 8," but she double-counted some of the squares. She then turned the prism 90 degrees so that she saw another face and restarted counting on that face. She counted "1, 2, 3, 4, 5, 6" on that face and then, rotating the prism another 90 degrees, counted "1, 2, 3, 4, 5, 6" on the next face. With each turn, she counted the total number of squares on that face, double counting the cubes at each edge. We saw this response as evidence of thinking at the Primitive 3-D Array Counter level because she demonstrated partial understanding of cubes as filling space: She made the choice to count when we asked her to compare the volumes but had difficulty maintaining an accurate count, focusing on counting squares on faces, resulting in the double counting of cubes.

During pre-K and kindergarten, children transitioned from counting squares on the faces of a three-dimensional solid constructed of cubes to eventually counting one cube at a time in carefully structured and guided contexts. Counting squares on faces often caused children to report an incorrect count either because they missed cubes (e.g., did not account for cubes on a top or bottom layer or cubes hidden in other layers) or because they double-counted cubes (e.g., counted the three visible squares of a corner cube separately).

As another example, in the spring of pre-K, we again asked children to organize the shapes in Figure 8.6 from the smallest number of blocks to the largest number of blocks, a task similar to the one on the initial assessment. In the S-shape and the L-shape, the two shapes that had only a single layer of cubes, Marina counted accurately. In the shapes that had more than one layer of cubes, however, she miscounted, pointing to and counting only the top of each cube. She apparently only counted the faces that were in her line of sight.

Figure 8.6. Sorting-solids-by-volume task from spring of pre-K.

Our evidence indicated that children progressed into the Primitive 3-D Array Counter level after solidly demonstrating the ability to compare volumes directly. In addition, thinking at this level was evident before thinking at the Capacity Relater and Repeater level for all children; thus, we determined that we had accurately placed the Primitive 3-D Array Counter level within the developmental progression of the LT for volume measurement.

Capacity Relater and Repeater

When children transitioned from the Primitive 3-D Array Counter level to the Capacity Relater and Repeater level, they became more aware of the relationships among volume, number of cubes, and three-dimensional space. In contrast to children at the Primitive 3-D Array Counter level, they had developed an understanding of the need for equal-sized units when determining a volume and they were able to keep track of those units. At the Capacity Relater and Repeater level, children use simple units to fill containers with accurate counting and measure volume by repeated use of the unit. At that level, children can iterate a single unit to measure the volume of an object and recognize at least intuitively that they should use identical units, but they may not initially appreciate the need for identical units in every situation. In addition, with teaching, a child at this level can relate the size and number of units and understands that increasing the size of the unit results in fewer units being needed.

We had previously observed the parallel development of capacity understanding and spatial structuring. On the basis of this observation, even though the initial developmental progression focused solely on capacity relating and repeating, we decided not to limit our tasks involving relating and repeating units to units of capacity. Instead, we included tasks in our teaching episodes that focused not only on filling but also on packing, with the expectation that we would integrate these in the revised developmental progression of the LT for volume measurement.

During the initial assessment in pre-K, Edith showed an initial foray into the Capacity Relater and Repeater level on the task depicted in Figure 8.1. She took the cube, held it inside the top of the box, and iterated along the opening, counting "1, 2, 3, 4." Although Edith did not display Capacity Relater and Repeater as her dominant level of thinking on this task because she did not realize that there would be a second layer, she clearly recognized a unit, maintained that unit mentally, and iterated to fill the space. Edith more consistently demonstrated Capacity Relater and Repeater reasoning in the teaching episode during the spring of Grade 1. In that teaching episode, we tasked her with finding the volume of a $3 \times 3 \times 4$ clear plastic box, given one $1 \times 1 \times 1$ wooden cube. She began by iterating the cube in rows along the bottom of the container, then iterated it up and down each of the sides, and finally iterated in the interior (see Figure 8.7). Although she miscounted slightly and arrived at a total of 34, Edith made a clear attempt to maintain equal units while she iterated throughout the container. We see this as consistent with thinking at the Capacity Relater and Repeater level.

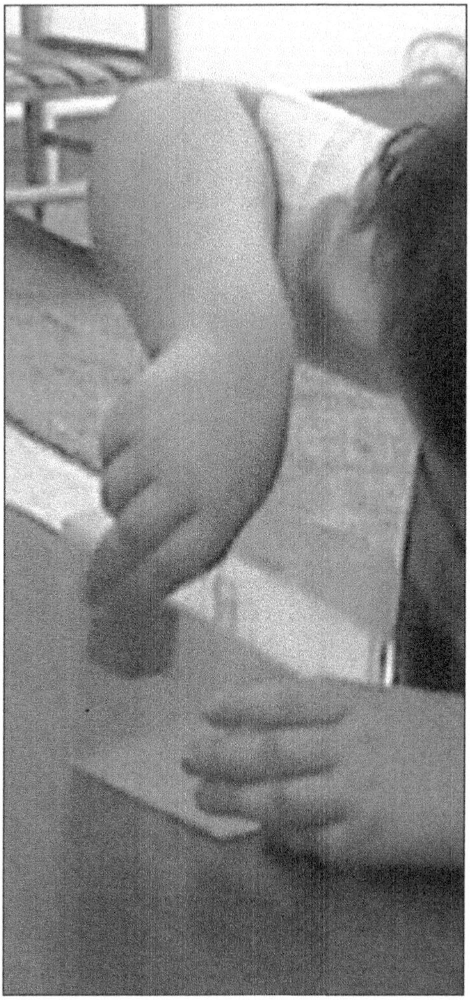

Figure 8.7. Edith iterating a cube to find the volume of a container.

Also in spring of Grade 1, we presented Edith with a hypothetical situation in which two children had measured the volume of the same box, but each had used a different unit. Edith stated that one child used a large cube whereas the other child used a smaller cube, and so they arrived at different counts. Edith, however, quickly pointed out that it was not correct that the two children arrived at different answers. She explained that we needed to know the correct volume, so each child must use cubes that are the same size. Further, she clarified that the child who used the larger cube would find a smaller number than the child who used the smaller cube. We interpreted her reasoning as evidence of thinking at the Capacity Relater and Repeater level.

Other children also began to demonstrate thinking at the Capacity Relater and Repeater level toward the end of Grade 1. In the spring of Grade 1, we gave Robert and Lia a small 1 × 1 × 1 cube scoop and asked each to estimate the number of scoops necessary to completely fill a 2 × 2 × 3 container. Both children estimated inaccurately and thought that four scoops would be necessary; however, when we asked them to fill the container, both Robert and Lia accurately counted while they poured each correctly filled scoop into the larger container until the container was full, to arrive at a total of 12 scoops. In addition, when we showed them a larger scoop, both children articulated that fewer of the larger scoops would fill the container. The children's use of simple units to fill a container with accurate counting and the recognition that they needed fewer larger units than smaller units to fill the given container provided evidence characteristic of thinking at the Capacity Relater and Repeater level.

The children in the Northeast cohort did not appear to remain at the Capacity Relater and Repeater level for very long because the tasks presented here occurred during the last teaching episode in the spring of the third year of the study. Robert, Lia, and Edith all exhibited dominantly Capacity Relater and Repeater level thinking in the spring of the third year of the study but demonstrated thinking at the Partial 3-D Structurer level by the beginning of the fourth year. Likewise, during the first teaching episode in the fall of Grade 2, most of the children already exhibited thinking at the Partial 3-D Structurer level.

Partial 3-D Structurer

Children at the Capacity Relater and Repeater level are able to determine the volume of an object by repeatedly filling with a single unit. At the Partial 3-D Structurer level, they begin to develop the ability to visualize and keep track of cubes, even those that they may not have directly handled (e.g., internal cubes) because children demonstrating thinking at the Partial 3-D Structurer level are able to build, maintain, and manipulate a mental image of a unit of units (i.e., a row or column). However, they do not yet use layers or multiplicative thinking. Instead, the children's counting behaviors include additive thinking, such as using the row and column structure along with skip counting, to find a total.

Lia did not begin to move into the Partial 3-D Structurer level until the final year of the study; and even then, she was unable to recognize internal cubes without strong guidance from the researcher. However, Robert and Edith showed evidence of thinking at the Partial 3-D Structurer level before Grade 1.

Robert made an initial venture into the Partial 3-D Structurer level in the spring of kindergarten. When we asked him to determine the volume of a 2 × 2 × 3 rectangular prism constructed from the cubes, he played with the solid for several seconds. While Robert studied the prism, he turned it around on the table several times before picking it up and attempting to count around the bottom layer. He then turned the prism in his hands when he reached the corners in an attempt to keep track of the cubes that he had already counted. He restarted his counting a couple of times, double-counted at corners, and arrived at a total of nine. Then, unprompted, Robert laid the prism on the table, counted the six cubes in the top layer and flipped the prism over. Finally, he counted the six cubes in the bottom layer and arrived at the correct total of 12.

Although he began to show more advanced structuring as well as recognition of cubes as filled space, that task did not include any internal cubes. Therefore, we could not claim that Robert was dominantly at the Partial 3-D Structurer level. In fact, it was not until a teaching episode during the fall of Grade 1 that Robert exhibited thinking at the Partial 3-D Structurer level by accounting for internal cubes. We asked Robert to find the volume of a 3 × 3 × 4 rectangular prism that we had constructed by using multilink cubes. He immediately recognized that nine cubes were on the top of the prism. He then followed up by saying, "There are nine in each of them, so nine plus nine plus nine plus nine." Although we did not ask Robert to calculate a total, his thinking was clearly additive, and he was clearly accounting for internal cubes even though he could not see them. We therefore saw his thinking as indicative of the Partial 3-D Structurer level.

In the spring of kindergarten, Edith showed initial movement into the Partial 3-D Structurer level when we asked her to find the volume of a 2 × 2 × 3 rectangular prism made of small wooden cubes. In contrast to Robert, who picked up the prism, Edith left the prism on the table with a 2 × 3 face positioned in front of her. She proceeded to count two cubes in each column, starting in the front-left corner, and continued to the back of the prism without moving it. She then counted across the front, correctly determining the volume as 12. This three-dimensional figure did not contain interior cubes, but Edith counted systematically and indicated that she recognized the "hidden" cubes because she counted cubes on the bottom layer of the back face that she could not see. We interpreted this process as evidence of thinking at the Partial 3-D Structurer level.

Edith continued to demonstrate thinking at the Partial 3-D Structurer level into Grade 2. In the same task previously described for Robert, we presented Edith with the 4 × 3 × 3 rectangular prism constructed of cubes. She held the prism so that she looked at a 4 × 3 face with the 3 × 3 face on the top and began counting on the front face, counting four in each column. Edith then turned the cube and counted the remaining columns on the other faces, being careful to avoid double counting at the corners. When she arrived at the middle column (the one with hidden cubes), Edith pointed to the middle cube on the top face, then pointed and counted down the side. In addition, when the researcher asked Edith to clarify her thinking, she stated that she knew that the squares at the corners on different faces of the prism belonged to the same cube, which led her to count the squares only once. She went on to say that for the middle column, she used the number of cubes in the column on the side that she could see to guide her counting of the hidden cubes. When the researcher asked how many cubes were in the center column, Edith stated that there were four. Again, we interpreted her use of columns as a structure and recognition of internal cubes as indicative of thinking at the Partial 3-D Structurer level.

More Sophisticated Levels Than Partial 3-D Structurer

None of the children in the Northeast cohort demonstrated consistent thinking at levels more sophisticated than Partial 3-D Structurer. We next transition to further explanation of the qualitative data already presented and leave the discussion of higher levels of the developmental progression within the LT for volume measurement to the following chapter, which focuses on the children in the Midwest cohort.

LONGITUDINAL GROWTH CHARTS

To help summarize the preceding qualitative data, we present individual longitudinal growth charts for three of the children in the Northeast cohort—Robert, Edith, and Lia—for whom we have 4 years of data. In each chart, vertical lines represent the levels of thinking that the child demonstrated during each teaching episode involving volume tasks over the 4-year period. In addition, a thicker section of the line in each chart represents what we identified as the dominant level of thinking that the child demonstrated during that teaching episode (see Chapter 2 for a more detailed description of the growth-chart components). Along with each child's chart, we give a short summary and interpretation of what the growth chart presents.

Robert's Growth

In pre-K, Robert demonstrated thinking predominantly at the Capacity Direct Comparer level, although we saw evidence that his thinking reached into levels as high as the Partial 3-D Structurer level (see Figure 8.8). When Robert moved into kindergarten, he used dominantly Capacity Direct Comparer level thinking but progressed up to the Primitive 3-D Array Counter level by the end of the school year, indicative of development in his spatial structuring ability. When Robert entered Grade 1, we recognized that his spatial structuring ability was already developed to the point that his dominant level of thinking was at the Partial 3-D Structurer level. Further, Robert continued to display Partial 3-D Structurer as his dominant level of thinking for the remainder of our study (through the end of Grade 2).

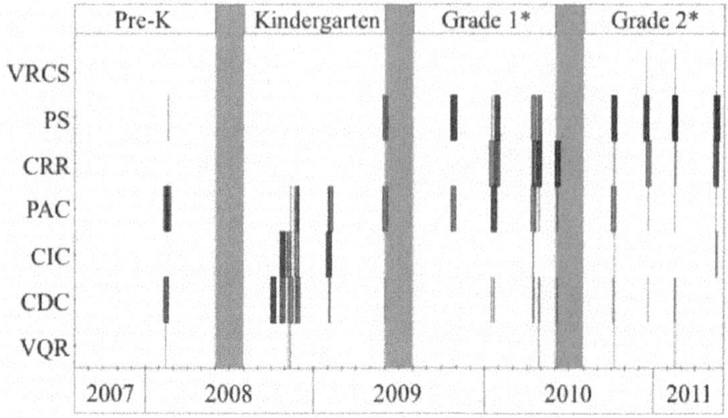

Figure 8.8. Robert's longitudinal growth chart. VQR = Volume Quantity Recognizer, CDC = Capacity Direct Comparer, CIC = Capacity Indirect Comparer, PAC = Primary 3-D Array Counter, CRR = Capacity Relater and Repeater, PS = Partial 3-D Structurer, VRCS = 3-D Row and Column Structurer.

Edith's Growth

Edith used dominantly Primitive 3-D Array Counter level thinking during pre-K (see Figure 8.9), indicating that her spatial structuring ability had already begun to develop. During kindergarten, her spatial structuring ability continued to develop while she transitioned to using predominantly Partial 3-D Structurer level thinking by the end of the school year. Although one might interpret the figure as indicating that her thinking fell back for most of kindergarten, we assert that this apparent fallback occurred because most tasks during that year focused on capacity rather than on spatial structuring. The tasks did not address the levels of the developmental progression for volume measurement that focus on spatial structuring and limited the thinking that she could demonstrate. Edith displayed Partial 3-D Structurer as her dominant level of thinking throughout Grade 1 and Grade 2, although toward the end of Grade 2, she was beginning to reach into the 3-D Row and Column Structurer level. On three occasions in Grade 1, during teaching episodes that focused specifically on capacity, the dominant level that we observed for Edith was Capacity Relater and Repeater. During those teaching episodes, similar to the apparent fallback in kindergarten, the highest level that she could demonstrate was Capacity Relater and Repeater because all levels beyond that in the initial developmental progression focused on spatial structuring. Even though we identified Capacity Relater and Repeater as Edith's dominant level of thinking because she had demonstrated higher levels of thinking from the end of kindergarten, we did not take this to be evidence that she had moved to lower levels of thinking. Rather, we believe that the tasks presented to her on these occasions constrained her thinking.

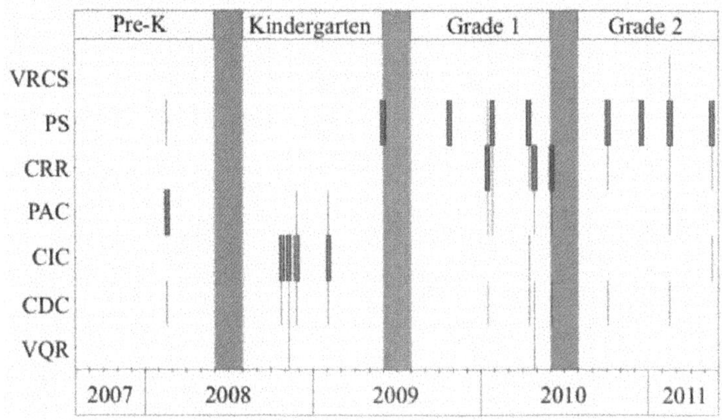

Figure 8.9. Edith's longitudinal growth chart. VQR = Volume Quantity Recognizer, CDC = Capacity Direct Comparer, CIC = Capacity Indirect Comparer, PAC = Primary 3-D Array Counter, CRR = Capacity Relater and Repeater, PS = Partial 3-D Structurer, VRCS = 3-D Row and Column Structurer.

Lia's Growth

Lia, like Edith, demonstrated Primitive 3-D Array Counter as her dominant level of thinking in pre-K (see Figure 8.10). In contrast to Edith, however, Lia continued to maintain that level of thinking throughout the second year and into the first half of the third year of the study before her thinking progressed to the Capacity Relater and Repeater level. During the final year of the study, she demonstrated inconsistent thinking, sometimes dominantly at the Primitive 3-D Array Counter level, while at other times her thinking was at the Capacity Relater and Repeater level or even at the Partial 3-D Structurer level.

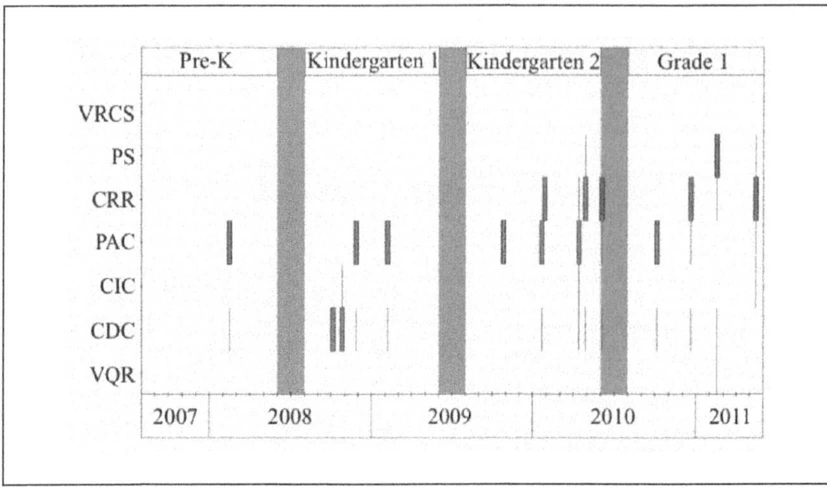

Figure 8.10. Lia's longitudinal growth chart. VQR = Volume Quantity Recognizer, CDC = Capacity Direct Comparer, CIC = Capacity Indirect Comparer, PAC = Primitive 3-D Array Counter, CRR = Capacity Relater and Repeater, PS = Partial 3-D Structurer, VRCS = 3-D Row and Column Structurer.

DISCUSSION AND SUMMARY

Figure 8.11 presents a composite growth chart for Robert, Edith, and Lia. We have based several observations regarding children's thinking about and understanding of volume, as well as the developmental progression of the LT for volume measurement itself, on our examination of this chart and the preceding charts in this chapter.

First, overall growth occurred in the dominant level of children's thinking about volume while they moved from pre-K through Grade 2. That growth is evidence that the developmental progression is consistent overall with behaviors that the children in our study demonstrated. This qualitative evidence through longitudinal data supports the developmental progression of the initial LT for volume measurement.

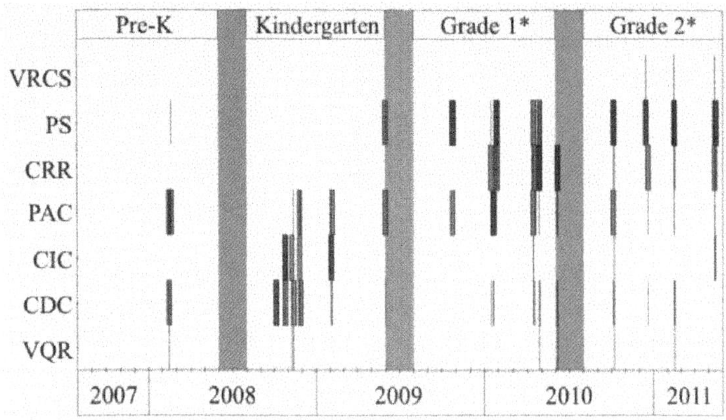

Figure 8.11. Composite growth chart for three children. Asterisks on Grade 1 and Grade 2 headings indicate that the Grade 1 column reports Lia's data from her second year of kindergarten, the third year of our study, and the Grade 2 column reports Lia's Grade 1 data. VQR = Volume Quantity Recognizer, CDC = Capacity Direct Comparer, CIC = Capacity Indirect Comparer, PAC = Primitive 3-D Array Counter, CRR = Capacity Relater and Repeater, PS = Partial 3-D Structurer, VRCS = 3-D Row and Column Structurer.

Second, the kindergarten year seemed to be a year when volume understanding was developing, as indicated by the wide distribution and variability of the dominant levels of thinking that these children demonstrated during that year. This result may be evidence that children's spatial structuring ability was developing during that time, but it also may highlight that the initial developmental progression of the LT for volume measurement presented volume schemes inconsistently and separately across levels. That is, the developmental progression began with an initial level that focused on understanding volume as a quantity, then moved to focus on capacity for the next two levels before shifting to spatial structuring.

Third, both Robert and Edith exhibited a shift to using Partial 3-D Structurer as their dominant level of thinking in the third year of the study, during Grade 1. Lia, however, did not demonstrate this same shift, perhaps because she had repeated kindergarten. Although repeating kindergarten would not have affected her development, it did affect her curriculum, because she was not exposed to the Grade 1 curriculum until the final year of the study. Lia too demonstrated the most inconsistency in her dominant level of thinking during the final year of the study. Further examination of the videotapes of her teaching episodes revealed that, overall, her spatial structuring ability was fairly stable and well-developed for less complex tasks; however, she often struggled to maintain focus as the tasks became more complex.

By the end of the study, all three children demonstrated an understanding and fluency with units of units; that is, they could build and maintain mental images of rows or columns and could use additive reasoning to find the volume of an object. One child, Edith, had begun to visualize layers, but she resorted to rows or columns and additive thinking when we asked her to determine volume. The dominant level of thinking most representative of children in our study at the end of Grade 2 was the Partial 3-D Structurer level.

The next chapter discusses the volume understanding of the children in the Midwest cohort, which includes some overlap of levels that we have previously discussed in connection with the younger children in the Northeast cohort. We also describe the children's progression toward more sophisticated levels of thinking.

Chapter 9

Volume in the Middle and Later Elementary Grades

Craig J. Cullen
Illinois State University
Jeffrey E. Barrett
Illinois State University
Melike Kara
Towson University
Cheryl L. Eames
Southern Illinois University, Edwardsville
Amanda L. Miller
Illinois State University
David Klanderman
Trinity Christian College

The previous chapter highlighted the longitudinal growth of children's understanding of volume measurement over a 4-year period, spanning from pre-K to Grade 2. We next discuss more sophisticated levels of the developmental progression of the learning trajectory (LT) for volume measurement by examining the work of children throughout a comparable 4-year period, spanning Grade 2 to Grade 5. This chapter has two main purposes. First, we want to describe and illustrate the more advanced levels of the developmental progression of the LT for volume measurement. Second, we want to follow one particular child, Abby, while she moved through the levels in the developmental progression. To accomplish this goal, we will share Abby's work that was consistent with each of the relevant levels when we explain each level. This child's work provides a central orientation to the sequence of levels because her volume measurement development was relatively steady. In addition, her growth spanned all the levels that this chapter discusses, from Primitive 3-D Array Counter to 3-D Array Structurer.

To furnish a wider perspective of children's thinking, we supplement the narrative from level to level by including examples for each level from one or more of the other four children in our Midwest cohort. After describing and illustrating the levels, we zoom out and present longitudinal growth charts for each of those five children. These charts attempt to provide an overview that indicates how each of the five children in the Midwest cohort moved through the levels. Finally, we present a composite growth chart and discuss overall patterns from the longitudinal analysis of this cohort of children.

CHECKING THE INITIAL LT FOR VOLUME MEASUREMENT: QUALITATIVE EVIDENCE

Primitive 3-D Array Counter

Although children at the Capacity Direct Comparer and Capacity Indirect Comparer levels produce mainly qualitative comparisons (e.g., "This container has a larger capacity"), children at the Primitive 3-D Array Counter level begin to transition to quantifying the volume and capacity of three-dimensional objects. Children at the Primitive 3-D Array Counter level demonstrate a partial understanding of cubes as filling a space, but they do not attend to interior cubes. At this level, children often make mistakes such as counting the visible cubes (or squares) on the faces of a rectangular prism or double counting the cubes on the corners or edges. The previous chapter included accounts of children's thinking that described the levels for qualitative comparisons (i.e., Capacity Direct Comparer and Capacity Indirect Comparer), and for quantitative comparisons at less sophisticated levels, including the Primitive 3-D Array Counter level. Because the children in the Midwest cohort entered the study in Grade 2, we expected some overlap with levels achieved by the children in the Northeast cohort during earlier grades (mainly Grade 1).

Abby primarily exhibited thinking at the Primitive 3-D Array Counter level at the outset of this study during Grade 2. We did not focus exclusively on volume topics in a teaching episode until the middle of the Grade 3 year. During that teaching episode, the researcher presented Abby with a two-dimensional representation of two gridded rectangular prisms (a 3 × 2 × 2 prism and a 4 × 2 × 3 prism; see Figure 9.1) and asked her to compare the prisms by volume.

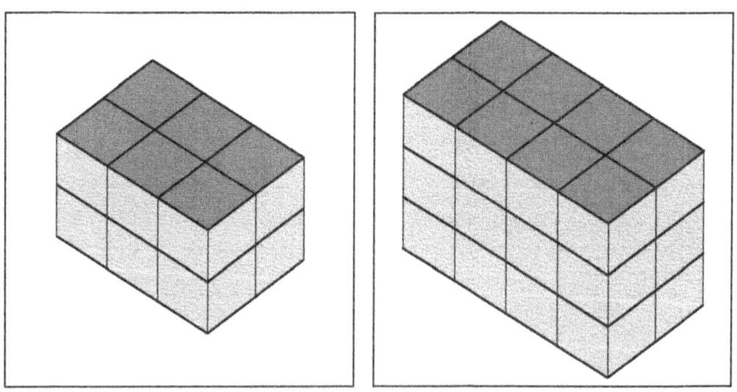

Figure 9.1. Two gridded rectangular prisms, a 3 × 2 × 2 prism and a 4 × 2 × 3 prism.

Abby counted only the visible cubes on three visible faces and said that the volume of the 3 × 2 × 2 prism was 16 and that the volume of the 4 × 2 × 3 prism was 26, which is incorrect. Her attention to the visible cubes on the faces illustrated thinking at the Primitive 3-D Array Counter level. The researcher asked Abby to check her answer by building the figures and gave her a pile of wooden one-inch cubes. Abby correctly built the 3 × 2 × 2 prism; however, she incorrectly built the 4 × 2 × 3 prism by building three layers of 12 instead of three layers of eight. After more questioning from the researcher, Abby recognized this mistake. Abby's counting the squares on the visible faces of the prisms is consistent with the Primitive 3-D Array Counter level. This teaching episode was the only one in which Abby predominantly used thinking at the Primitive 3-D Array Counter level in response to tasks.

In contrast to Abby's short-lived demonstration of the Primitive 3-D Array Counter level of thinking, Drew demonstrated this type of thinking during approximately half of the volume interviews. For example, during the winter of Grade 4, we gave Drew a two-dimensional representation of a gridded 4 × 6 × 3 rectangular prism shown next to a unit cube (see Figure 9.2). The researcher asked, "How many of these blocks [pointing to the unit cube] would you need to make that [pointing to the 4 × 6 × 3 prism]?"

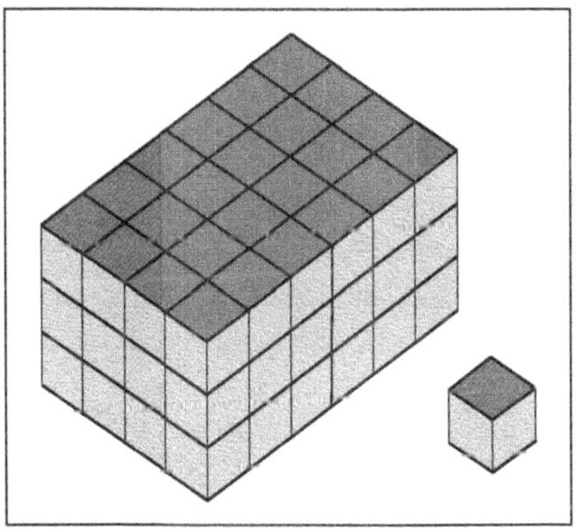

Figure 9.2. Gridded 4 × 6 × 3 rectangular prism.

Drew initially counted 18 unit cubes while touching each square on the visible 3 × 6 face, which he called a "wall." He explained that there was another "wall" of 18, a 3 × 6 face that was not visible in the picture. He then skip-counted by threes to count 12 on the visible 4 × 3 wall and explained that another identical wall of 12, a 4 ×3 face, was not visible in the picture. He counted 24 while touching each square on the top face, and then added 18 + 18 + 12 + 12 + 24 to obtain his answer of 84. When the researcher asked him about his strategy, however, Drew pointed to the four edge cubes and explained that his count of 84 was too high "because this right here is all the same block." He then concluded that it should be 60 because he "shouldn't have counted the top." Although he did not correctly compensate for his overcounting, he did notice that some overcounting had occurred in the task.

At times during this teaching episode, Drew appeared to be counting cubes on visible and nonvisible faces; however, he touched and counted both squares from the corner cubes. This action made us question whether Drew attended to cubes or was merely counting squares on the faces of the prism. Drew's actions of counting faces, double counting cubes at the corners, and not counting internal cubes indicated that he was reasoning at the Primitive 3-D Array Counter level.

Although differences existed in the ways that Drew and Abby completed these volume tasks, neither attended to internal cubes and both double-counted corner cubes. These behaviors were all consistent with the Primitive 3-D Array Counter level. Drew was closer to transitioning from Primitive 3-D Array Counter than Abby because he attended to cubes that were not visible and because he recognized that some squares on different faces of the prism belonged to the same cube. The researchers presented these tasks as two-dimensional representations of three-dimensional objects, thereby increasing the demand of volume-measurement tasks, Examining whether children at this Primitive 3-D Array Counter level may be able to advance a level by working directly with solid boxes in three dimensions rather than having to interpret the three-dimensional object when the researcher presents only two-dimensional images of it is a matter for further research.

Capacity Relater and Repeater

When children transition from the Primitive 3-D Array Counter level to the Capacity Relater and Repeater level, they become more aware of the relationships among volume, cubes that constitute parts of objects, and the space in which objects fit. In contrast to children at the Primitive 3-D Array Counter level, children at the Capacity Relater and Repeater level intend to fill space with cubes and keep track of those cubes, but they may still make mistakes with volume. Children at the Capacity Relater and Repeater level demonstrate the action of filling a container by repeatedly filling a unit and counting how many of the units are necessary to fill the container. They show the ability to iterate a mental unit along a perceptually available object. In addition, with appropriate instruction, they come to understand that they will need fewer large objects than small objects or units to fill a given container.

During the spring of Grade 3, the researcher asked Abby to compare the volume of two transparent rectangular prisms; Prism A was 1 in. × 1 in. × 2 in., and Prism B was 1 in. × 2 in. × 2 in. (see Figure 9.3). Abby put the prisms next to each other and correctly identified the larger prism. When the researcher asked her how much larger it was, she stated that the small prism was about half the size of the large prism and that two of the small prisms would fit into the large prism. Next, the researcher asked, "When you say this one [Prism B] is two, what is your unit?" After pausing for a while, Abby answered, "Two inches." She appeared to recognize that the smaller object being counted was itself 2 inches long, although she may have also meant that it was 2 cubic inches.

Understanding the intent and meaning for an answer such as "two inches" in this situation is demanding without further information. We found that always asking children whether their wording fit the units being measured was impractical, although it is an open question whether such careful and constant insistence on naming units might support conceptual shifts in reasoning and understanding. In practice, we look for clarification by context from previous and subsequent interactions.

Next, the researcher introduced some cubes with 1-inch side lengths, and Abby identified the objects as cubic inches. The researcher took two of the 1-inch cubes and asked Abby if two of these cubes would fit into the larger prism (see Figure 9.3). When Abby heard this question and saw the arrangement of two 1-inch cubes, she changed her answer about Prism B, saying that it would be four cubes. Next, the researcher used inch cubes to compose a 2 in. × 2 in. × 1 in. prism and placed it adjacent to the larger prism. The researcher asked Abby how many smaller prisms (i.e., Prism A) would be equal in volume to the prism built from the cubes (i.e., Prism B). She said that the answer would be two. When the researcher asked Abby how she arrived at that answer, Abby explained that two of the small prisms would be one large prism. The researcher then asked, "If the large block was two, what would the unit be?" Abby correctly stated that Prism A would be the unit. Next, the researcher asked, "If this is my unit [picking up Prism A], then what would the volume of this be [picking up Prism B]?" Abby paused and then stated that the small prism would be half. We believe that she did not answer the question that the researcher posed, but rather reversed the unit to the larger prism, Prism B, and described the size of Prism A. This idea is only a speculation on our part, however. In any case, Abby made a true statement about the relationship of the smaller prism to the larger one, even though she did not answer the question asked.

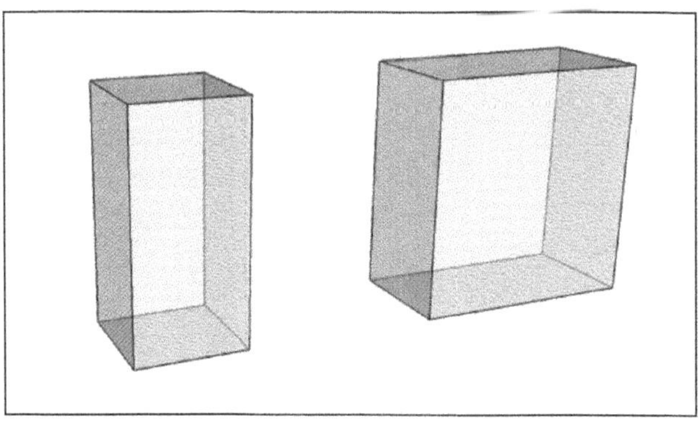

Prism A: 1 × 1 × 2 Prism B: 1 × 2 × 2

Figure 9.3. Comparing the volume of two prisms.

The researcher next opened the lids of the two prisms, gave Abby the larger prism, and asked her to fill that prism with enough water to exactly fill the smaller one. Abby first filled the larger prism completely with water and realized that she had much more water than she needed for filling the smaller prism. She poured out half the water and stated that this amount of water, which filled almost half of the prism, would fill the small block perfectly. The researcher asked again, if the large prism was one, how big the other prism was. Abby correctly said, "half." Last, the researcher asked if the large prism was one, what would be the volume of an inch cube, and Abby correctly said, "a quarter."

Abby did not struggle when the researcher asked her to find the volume of the large prism in terms of the small prism. She immediately said "two." However, when the researcher asked her to specify the unit (i.e., "The large prism is two what?"), Abby said that the unit was inches. We interpreted this response as an indication that her understanding of a volume unit was not fully developed. After some instructional discussion, Abby could use the small prism as a unit to compare to the larger one. Filling the large prism with water was enough to help her realize that the volume of the small prism would be half the volume of the large prism.

We concluded that Abby needed more tasks involving a unit that was larger than the object that she was measuring to promote further experience with ratios and greater understanding of ratios. When the researcher presented prisms in the ratios of 1:2, 1:4, 1:½, and 1:¼, Abby's responses suggested that she coped with arbitrary assignments of a unit, including situations in which the unit was larger than the object to be measured, although sometimes only within the context of pouring. We inferred that her thinking was consistent with the Capacity Relater and Repeater level. Anselm, Arielle, and Owen demonstrated similar responses to this task.

In contrast, Drew did not predict the correct answer on the subtask "If this [$1 \times 2 \times 2$ prism] is one, what would this [$1 \times 1 \times 2$ prism] be?" Instead, he responded, "maybe . . . one . . . one and a half." He did not relate units of different sizes. In this task, we saw varying levels of success with different solids that we had arbitrarily defined as a unit. Such flexibility of thinking about units in a given context continues to improve at the next level of the developmental progression, the Partial 3-D Structurer, when children begin to count collections of units as a way to increase the efficiency of their cube-counting strategies (see Barrett et al., 2011, for more information on these prism tasks and related findings).

Partial 3-D Structurer

Although children at the Capacity Relater and Repeater level are able to find the volume of an object by repeatedly filling that space with a single unit, children at the Partial 3-D Structurer level are beginning to develop the ability to imagine and keep track of cubes that they did not directly handle. They use more efficient ways of summing units to find volume as a quantity. Children thinking at this level have made a conceptual advance in their spatial structuring strategies and use this structure to guide their enumeration. They have developed more accurate strategies that allow them to account for groups of cubes.

When children initially reach the Partial 3-D Structurer level, they attempt to account for internal cubes (i.e., cubes without visible faces) when they determine the volume of an object but they may still count unsystematically. Over time, children at the Partial 3-D Structurer level develop counting strategies that are more systematic and that allow them to see a cube as both a unit and as a component of a row or column in a three-dimensional structure. Despite these advances in counting and spatial structuring strategies, children at the Partial 3-D Structurer level cannot yet attend to a layer (i.e., a unit of units of units—this ability would signal the 3-D Row and Column Structurer level of thinking).

All five children demonstrated thinking at the Partial 3-D Structurer level during at least one teaching episode in our study. One child, Drew, exhibited an approach that was at the Partial 3-D Structurer level just twice during the study; whereas another child, Anselm, demonstrated thinking at the Partial 3-D Structurer level on multiple occasions throughout the study. Other children showed this Partial 3-D Structurer approach intermittently. The first time that Abby appeared to use a Partial 3-D Structurer approach was during the spring of Grade 2. The researcher presented a two-dimensional representation of a $3 \times 2 \times 2$ gridded rectangular prism, provided 21 cubes, and asked, "Can you put some blocks together in a group that looks just the same as this drawing?" Abby built three rows of two cubes to create a horizontal layer of six and then built three more rows of two cubes to create a second layer of six on top of the first. Each time she built a row, she selected two cubes, one in each hand, placed them together, and then added them to the figure. Because Abby repeatedly selected two cubes at a time and created rows of cubes to form layers, she demonstrated that she was not just thinking about cubes as individual entities. Instead, she applied a row as a unit of units repeatedly, demonstrating that she was at least using Partial 3-D Structurer thinking. Because evidence existed that Abby was working on rows of cubes (units of units) but no indication existed that she was repeating an entire layer (a unit of units of units), we described her thinking as fitting the Partial 3-D Structurer level.

Owen exhibited thinking and behaviors at the Partial 3-D Structurer level once during Grade 2 and once during Grade 3. The last time that Owen exhibited thinking at the Partial 3-D Structurer level was during the spring of Grade 4, when the researcher presented a two-dimensional representation of a $4 \times 6 \times 3$ gridded rectangular prism (see Figure 9.2) and asked him to build the top layer. He built a $4 \times 6 \times 1$ layer by creating six rows of four cubes. When the researcher asked Owen how he had figured out the construction, Owen asserted that he built the top layer, noticed that there were two sets of 12, and subdivided his layer into two sets of $2 \times 6 \times 1$ arrays. The researcher then asked Owen to build the bottom layer, and he responded that the bottom layer would "probably be the same" but that he would need to build "a new one" for the middle layer.

The researcher then prompted Owen to build the middle layer. Again, he built six rows of four to create a layer of 24. This action demonstrated that Owen understood cubes as filling a space, that he accounted for inside and outside cubes, and that he structured space by using rows or columns of individual cubes. However, we did not find evidence that he was able to structure by using columns of rows or layers of individual cubes in more comprehensive structures. That kind of thinking would fit at the next level, 3-D Row and Column Structurer.

In contrast to Abby and Owen, Anselm used thinking and behaviors at the Partial 3-D Structurer level throughout the study; he used them in three teaching episodes, one each in Grades 2, 3, and 5. By Grade 4 he was reaching into higher levels of the LT for volume measurement, but he sometimes engaged in Partial 3-D Structurer thinking on tasks. For example, during the spring of Grade 4, the researcher presented Anselm with the outline of a box, as shown in Figure 9.4. The researcher asked him how many cubes he would need to fill the box.

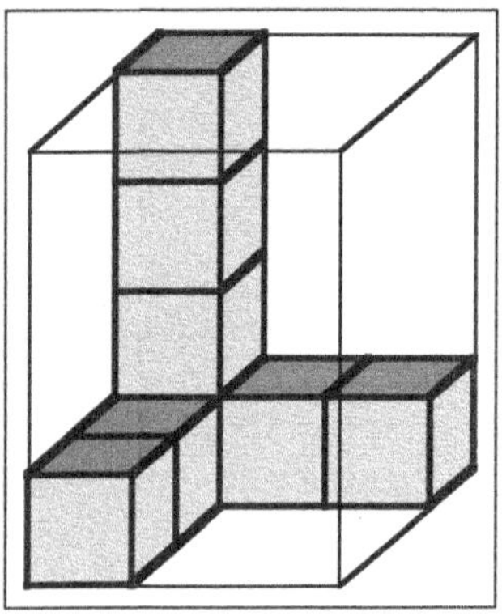

Figure 9.4. Box task image.

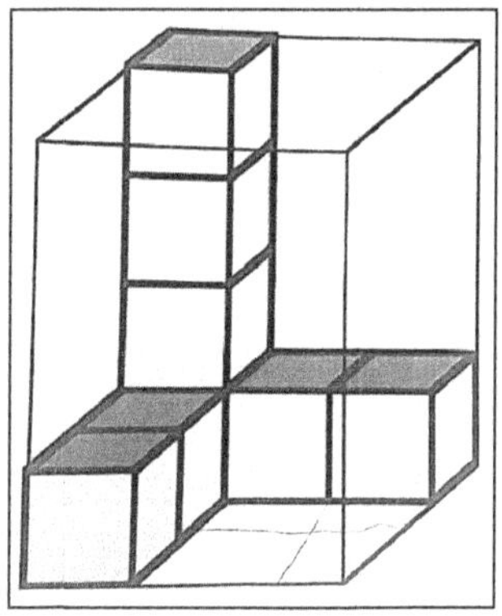

Figure 9.5. Anselm's drawing on the box task.

Before touching or drawing, Anselm answered 34. When the researcher asked Anselm how he obtained his answer, he explained that he counted by threes when cubes were already there (indicating the vertical cores along the back and left faces) but fours when a given portion of the prism did not display any cubes (i.e., an empty row). He then drew two segments on the base to indicate where the four vertical cores of four belonged, as indicated in Figure 9.5. When the researcher asked for clarification, Anselm correctly stated that 28 cubes were missing, 36 cubes would completely fill the box, and nine cubes would fill the bottom of the box. Interestingly, Anselm was the only child who was able to use thinking at the Partial 3-D Structurer level to resolve this task correctly. We believe that his approach here indicates that he was operating on a unit of units, the number of cubes in columns of a three-dimensional structure, and that he applied this unit of units repeatedly and exhaustively. His way of operating indicates that he was not yet using layers or multiplicative thinking, which are concepts that indicate the next higher level, the 3-D Row and Column Structurer level of the LT.

We presented empirical evidence of the Partial 3-D Structurer level for both the Northeast cohort (see Chapter 8) and the Midwest cohort to portray the full extent of this level of thinking. Because none of the children in the Northeast cohort exhibited thinking and behaviors more sophisticated than the Partial 3-D Structurer level, we can only draw on data from the Midwest site to describe children's transitioning from this level to more sophisticated levels. However, by comparing how children exhibited thinking and behaviors at the Partial 3-D Structurer level at the two sites, we note two additional findings. First, we observed repeated addition (see the Partial 3-D Structurer section of Chapter 8 for Robert's strategy) developing into multiplication (e.g., Abby's strategy described previously for the Partial 3-D Structurer level) during this level. Second, we adapted and designed a wider variety of tasks, including both two-dimensional and three-dimensional representations of volume tasks.

3-D Row and Column Structurer

When children are operating at the Partial 3-D Structurer level, they create and use rows or columns (i.e., unit of units) to count larger sets of units with increased efficiency. However, they are still working on how to demonstrate the creation and use of a layer (i.e., a unit of units of units). In contrast, children demonstrating thinking at the 3-D Row and Column Structurer level count or compute, row by column, the number of cubes in one row and then use addition or skip counting to determine the total for a layer. They build, maintain, and manipulate mental images of three-dimensional arrays, treating them as single entities and at the same time recognizing them as compositions of individual layers (i.e., units of units of units) containing individual rows (i.e., units of units) containing individual cubes (i.e., units).

Abby demonstrated this level of reasoning during Grade 4 and into Grade 5. During the spring semester of Grade 4, Abby encountered the image illustrated in Figure 9.6, and the researcher asked her how many cubes she would need to fill the box.

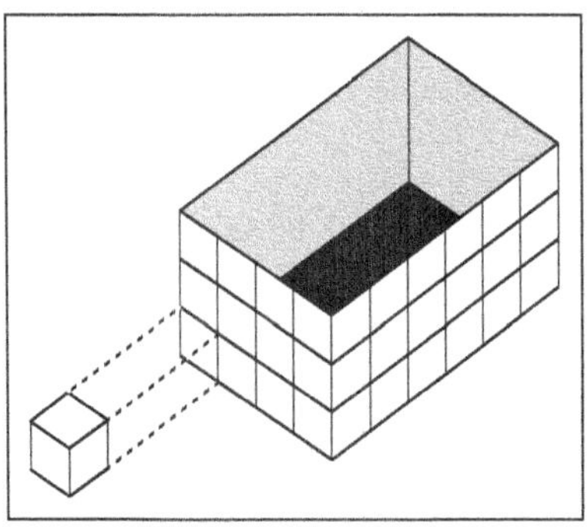

Figure 9.6. Gridded packing task.

Abby counted the number of cubes, one by one (not per row or column), on the front vertical face of 12 cubes, counted six along the top side of the exposed face, and then multiplied 12 by 6 to get 72 cubes, which was the correct result. She stated that she would have 12 in each layer (also correct). She explained, "So this side has 12 squares, so if . . . and then the length of this is 6, so I would just think that you would fill it up with 12 in each." Abby counted the number of cubes in one layer (i.e., a unit of units of units) and then multiplied by the number of layers to determine the total. Therefore, we consider her thinking to be consistent with the 3-D Row and Column Structurer level. Although Abby used multiplication, not enough evidence exists to allow us to conclude that she was at the 3-D Array Structurer level because the face of the prisms directly displayed the grid. She may have depended on the grid as scaffolding to answer the question, so we can only conclude that she was able to use such a structure and extend it when it was already provided.

Abby's reasoning on this task represented a substantial shift toward using multiplication to find the volume of rectangular prisms. Her multiplicative reasoning did not imply that she understood the creation of volume units as a transformation of three lengths. Instead she used multiplication to count collections (i.e., units of units or units of units of units) of a preselected unit, in this case the units formed with sets of cubes. Although we find clear evidence for the claim that she exhibited her understanding of multiplication as a basis to compute volume, we do not have adequate evidence to address the stronger claim that she understood how the multiplication of three lengths must produce volume. We believe that Abby created an understanding that three related lengths afforded the basis for volume: She could use Length 1 of a rectangular prism to compute the number of cubes that would fit along Length 1, creating a row or column; she could then use Length 2 to compute the number of these rows or columns that would fit along Length 2, creating a layer; finally, Length 3 indicates the number of these layers that fit along Length 3.

For example, when Abby was computing the volume of a 3 in. × 2 in. × 4 in. rectangular prism, instead of seeing the product of three lengths as transforming inches into square inches and then cubic inches, Abby operated only on cubes, and the dimensions became scaffolds for organizing the cubes. Three (one-inch cubes) were in a row, two rows of three (one-inch cubes) were in a layer, meaning that six (one-inch cubes) were in a layer, and four layers of six (one-inch cubes) were in the entire prism. This use of multiplication is consistent with the view that measurement is the process of quantifying an attribute of an object in reference to a chosen unit, which we take to be a unit cube (Sarama & Clements, 2009b). With this perspective, computing volume began with identifying a part of the object as a unit and then computing the number of those units necessary to be as large as the object itself. Thus, multiplication was a tool that Abby used to count the number of chosen units, in this case cubes, not an operation that transforms lengths into cubes. We have little evidence for this argument; we do not wish to argue that

Abby was not thinking in a more advanced way about transformation of linear units to a volume unit, but we instead are recognizing our lack of evidence to make a further claim.

Abby showed similar thinking at the 3-D Row and Column Structurer level by counting the number of cubes in a layer and then multiplying by the number of layers on two additional packing tasks. The first task used actual cubes and a box; whereas the researchers presented the second task as a picture. Both tasks were packing tasks even though the box representing the packing aspect in the second task had a grid on the outside. We believe that the grid might support children's reasoning with spatial structuring, but we still expected that children would need to examine ways of fitting the unitary packages (which were not always unit cubes; instead, some were rectangular prisms) into the larger box.

During the fall of Grade 4, the researchers presented Anselm with a packing task (see Figure 9.7). The task was part of a more complex task. During the first part of the task, the researchers asked Anselm how many of the blocks (i.e., unit cubes) he would need to fill the container.

Figure 9.7. Fill box given one layer and one column task.

Anselm looked at the figure and then counted by 20s to 100. He explained that he knew that 20 were in the bottom layer because four 4s total 16 and four more add to 20. Anselm counted by layers of 20 additively. He computed the number of cubes in a row and then used addition and skip counting to determine the total. This process suggests that he built, maintained, and manipulated a mental image of a layer and operated on this layer to complete the partially constructed three-dimensional array. Therefore, we placed his thinking at the 3-D Row and Column Structurer level.

3-D Array Structurer

Children at the 3-D Row and Column Structurer level demonstrate progress toward reasoning about volume in a way that models the volume formula. They organize their counting of cubes in a three-dimensional array by counting the number of cubes in a layer and then repeating that layer to find volume or capacity. However, children at the 3-D Array Structurer level take this reasoning one step further by connecting linear dimensions with the number of cubes. They compute the volume of rectangular prisms from their linear dimensions and explain how the multiplication of three lengths produces a measure of volume. At the 3-D Array Structurer level, children build, maintain, and manipulate abstract units (i.e., a unit of units of units of units); they can mentally decompose a rectangular prism into layers (i.e., units of units of units), which they can then decompose into rows (i.e., unit of units) of individual units of volume. The 3-D Array Structurer level is the highest level of thinking about volume measurement within the initial developmental progression of the LT for volume measurement.

During the spring of Grade 5, the researchers presented Abby with the same outline of a box that we had used with Anselm in Grade 4 (see Figure 9.4). The researcher asked her how many cubes altogether she would need to fill the box. Abby initially counted by vertical columns (like a core sample in the ground) and said that 32 was the answer. When the researcher asked how she had obtained her answer, she explained, "there's four columns and three rows, so then . . . just do . . . well there's three columns going this way and three columns going that way . . . so . . . 3 times 3 is 6 and 6 times 4 is . . . 20 . . . oh . . . 24." When the researcher asked about her multiplication of 3 times 3 to obtain a result of 6, Abby realized that she had made a computational error and said, "Oh! No! It's nine . . . then that would be 36." When the researcher asked how many blocks she would need to fill the bottom of the box, she correctly answered nine. The researcher then asked her to clarify why she had changed her answer from 32 to 36. Abby explained, "Because there's nine on the bottom . . . and 9 times 4 is 36." Abby's thinking was consistent with the 3-D Array Structurer level of the volume trajectory because she was able to complete the bottom layer of nine. She also iterated that layer and determined the volume of the rectangular prism.

By the spring of Grade 5, during the last teaching episode of the project, Abby used only approaches at the 3-D Array Structurer level to measure volume on the tasks that the researchers posed. For one of the tasks, the researcher presented her with an image of a gridded $2 \times 4 \times 2$ rectangular prism and a cube (see Figure 9.8) and provided her with a calculator.

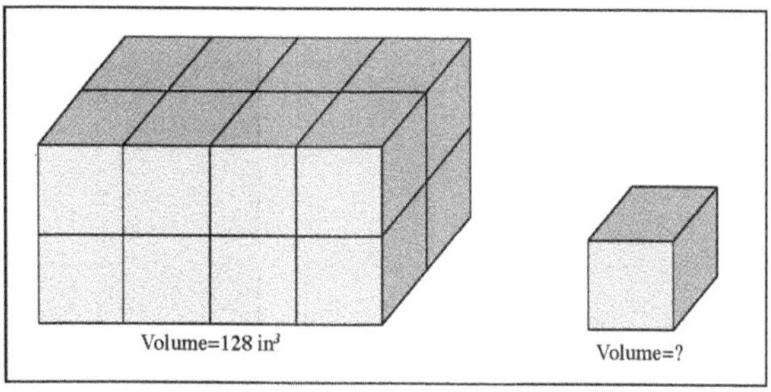

Figure 9.8. Find the volume of a single cube, given the rectangular prism with grid markings showing 16 cubes and a volume measure of 128 cubic inches.

The researcher told her that the prism had a volume of 128 cubic inches and asked her to find the volume of one cube. Abby used the calculator to divide 128 (the volume of the larger prism) by 16 (the number of cubes in the larger prism) to arrive at an answer of 8. She explained, "It is 16 cubes in the . . . rectangle . . . so . . . 8 cubic inches is the volume of each cube. The researcher then asked what the length, width, and height of the cube would be, and Abby pointed to three orthogonal sides saying, "This is two . . . , this is two . . . , and this is two." How Abby determined that each edge of the cube had a length of two is unclear. We conjecture that Abby was looking for three numbers that she could multiply to arrive at a total of 8. Although sophisticated ways exist to resolve this task, we believe that she used trial and error. The researcher then asked what the length, width, and height of the larger prism would be. Abby correctly labeled the length, width, and height of the larger prism as 4 inches, 8 inches, and 4 inches, respectively. Abby's thinking on this task is representative of the 3-D Array Structurer level because she was able to decompose a measure of volume into a product of three lengths.

Anselm began showing evidence of thinking at the 3-D Array Structurer level in the spring of Grade 3, which is much earlier than Abby's transition to using thinking and behaviors at the 3-D Array Structurer level. During the first teaching episode of the longitudinal study involving volume, the researcher gave Anselm a piece of paper with two-dimensional images of a 3 × 2 × 2 unit rectangular prism and a 4 × 2 × 3 unit rectangular prism (see Figure 9.1) and asked him to compare the difference in the volumes of the two prisms.

Anselm said, "about 12 more blocks." When the researcher asked how he knew, he said that he was counting by twos. He then counted on the top layer of the 4 × 2 × 3 unit rectangular prism, saying "2, 4, 6, 8" and then along the rightmost vertical 3 × 2 unit face, skipping the top row of two, saying "10, 12." He then explained that the smaller prism (i.e., the 3 × 2 × 2) would fit in what remained. Anselm's thinking is consistent with the 3-D Array Structurer level because he mentally decomposed a larger three-dimensional array structure into a smaller three-dimensional array structure to compare the prisms by volume.

Figure 9.9. Fill the box given one partial layer and one column task.

During the fall of Grade 4, as an extension of a previous task (see Figure 9.7), the researcher added cubes to one row and one column of the bottom layer (see Figure 9.9) and asked Anselm how many of the blocks he would need to fill the box, Anselm counted by fives to 25. The researcher asked where the 25 would be. Anselm pointed at the additional five cubes and said that he was figuring out "how many of these up it would be" to indicate that the row of five cubes became a partial vertical layer of 25. He then pointed at the extended row of five and said "here's another a hundred. So right now we're at 200." At that point, he had built a mental unit of units of units. He then used this mental unit of units of units to identify four collections of 100 cubes and determine that he would need 400 cubes to fill the box outlined with this partially constructed three-dimensional array.

In this teaching episode, Anselm built, maintained, and manipulated a mental image of a three-dimensional array of 100 cubes by building a row (i.e., a unit of units), which he then mentally iterated to build a layer (i.e., a unit of units of units), which he then mentally iterated to build a three-dimensional array (i.e., a unit of units of units of units), which he then mentally iterated to arrive at his final answer of 400 cubes. These actions, building up and using a three-dimensional array, are consistent with the 3-D Array Structurer level.

LONGITUDINAL GROWTH CHARTS

We next provide longitudinal growth charts for each of five children in the Midwest cohort. Each chart provides the dominant level of thinking (shown as a shaded rectangle) during a single interview. Unshaded rectangles or line segments are indications that the child used additional levels of thinking or behaviors during an interview but that we did not consider them dominant (see Chapter 2, page 45, for a more detailed description of the growth-chart components). For each child, we provide a brief interpretation of the growth chart.

Abby's Growth

In Grade 2, Abby used dominantly Primitive 3-D Array Counter level thinking and behaviors (see Figure 9.10). During Grade 3, she demonstrated Capacity Relater and Repeater as her dominant level of thinking. By Grade 4 she had shifted to using thinking at the 3-D Row and Column Structurer level, although she exhibited some variability (reverting back to thinking at the Partial 3-D Structurer level, the Capacity Relater and Repeater level, and the Primitive 3-D Array Counter level on some tasks). By the end of Grade 5, she had abandoned less sophisticated levels of thinking and transitioned to using 3-D Array Structurer as her dominant level of thinking.

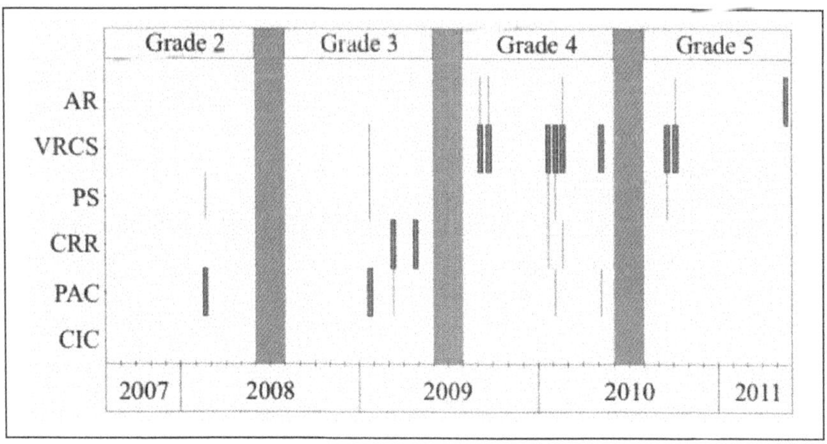

Figure 9.10. Abby's longitudinal growth chart.
CIC = Capacity Indirect Comparer, PAC = Primitive 3-D Array Counter,
CRR = Capacity Relater and Repeater, PS = Partial 3-D Structurer,
VRCS = 3-D Row and Column Structurer, AR = 3-D Array Structurer.

Owen's Growth

On the initial assessment during Grade 2, Owen exhibited dominantly Partial 3-D Structurer level thinking (see Figure 9.11). In Grade 3, he showed strong evidence of using thinking at the Capacity Relater and Repeater level and the Partial 3-D Structurer level during three teaching episodes that focused on volume measurement. During Grade 4, he made a clear shift to primarily using thinking at the 3-D Row and Column Structurer level, which he maintained into Grade 5. By the final teaching episode of the project, he showed evidence of thinking at the 3-D Array Structurer level on at least some tasks.

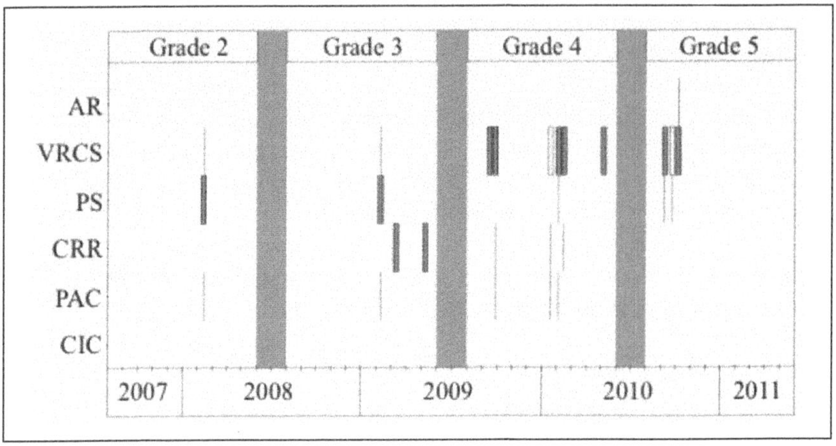

Figure 9.11. Owen's longitudinal growth chart.
CIC = Capacity Indirect Comparer, PAC = Primitive 3-D Array Counter, CRR = Capacity Relater and Repeater, PS = Partial 3-D Structurer, VRCS = 3-D Row and Column Structurer, AR = 3-D Array Structurer.

Arielle's Growth

Arielle exhibited Primitive 3-D Array Counter as her dominant level of thinking in Grade 2 and maintained steady growth through the levels of the developmental progression throughout the project (see Figure 9.12). She tended to exhibit thinking at the Capacity Relater and Repeater level more often by the end of Grade 3, moving toward dominantly Partial 3-D Structurer level thinking by the spring of Grade 4 and to dominantly 3-D Row and Column Structurer level thinking by the end of Grade 4. Arielle maintained her use of dominantly 3-D Row and Column Structurer level thinking during Grade 5.

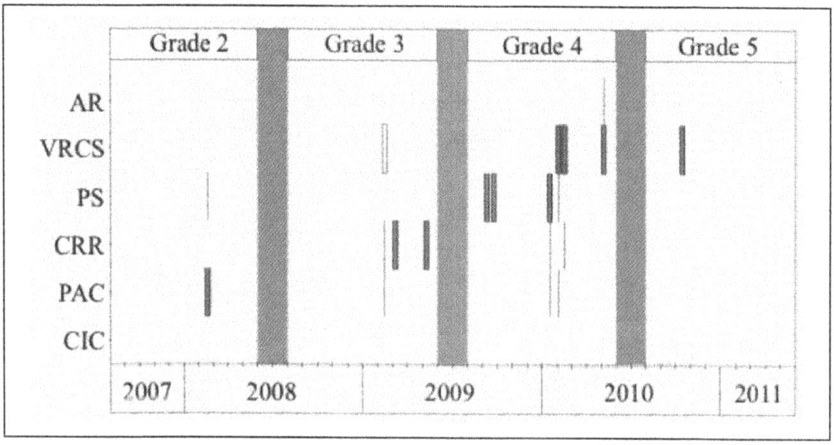

Figure 9.12. Arielle's longitudinal growth chart.
CIC = Capacity Indirect Comparer, PAC = Primitive 3-D Array Counter, CRR = Capacity Relater and Repeater, PS = Partial 3-D Structurer, VRCS = 3-D Row and Column Structurer, AR = 3-D Array Structurer.

Anselm's Growth

On the initial assessment during Grade 2, Anselm was consistently reasoning at the Partial 3-D Structurer level (see Figure 9.13). He exhibited more variability in his level of thinking in Grade 3, using a mixture of Capacity Relater and Repeater, Partial 3-D Structurer, and 3-D Row and Column Structurer levels of thinking across three teaching episodes. By Grade 4, however, he consistently showed 3-D Row and Column Structurer thinking, with hints toward a transition to the 3-D Array Structurer level. Other than one fall back to using dominantly Partial 3-D Structurer level thinking and behaviors early during Grade 5, Anselm maintained his use of thinking and behaviors at the 3-D Row and Column Structurer level and the 3-D Array Structurer level for measuring volume until the end of the project.

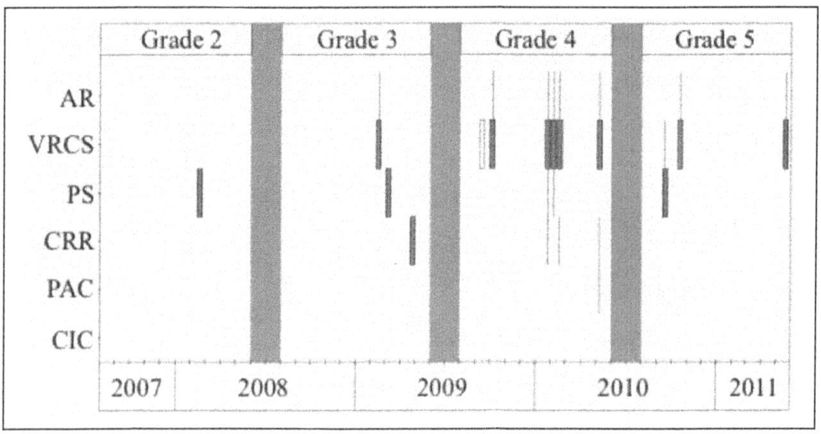

Figure 9.13. Anselm's longitudinal growth chart.
CIC = Capacity Indirect Comparer, PAC = Primitive 3-D Array Counter, CRR = Capacity Relater and Repeater, PS = Partial 3-D Structurer, VRCS = 3-D Row and Column Structurer, AR = 3-D Array Structurer.

Drew's Growth

Drew exhibited the most variability in his thinking on the variety of volume tasks posed. On the initial assessment during Grade 2, he showed Primitive 3-D Array Counter as his dominant level of thinking (see Figure 9.14). Depending on the aspect of volume or the representation of the three-dimensional object, he displayed a variety of thinking and behaviors, including thinking at the Capacity Relater and Repeater level, the Partial 3-D Structurer level, the 3-D Row and Column Structurer level, and even the 3-D Array Structurer level during Grades 3 and 4. His thinking and behaviors shifted with the task context more directly than any other cohort child. Our research team had limited opportunity to conduct further interviews with Drew during Grade 5, but we engaged him briefly on some volume tasks in early fall of that year. At that time, we followed up with him on the kinds of tasks with which he had struggled during Grade 4, and he exhibited the same kinds of struggles. On the final assessment during Grade 5, Drew used thinking at the Primitive 3-D Array Counter level and the Capacity Relater and Repeater level on some tasks, but he showed clear evidence of demonstrating Partial 3-D Structurer as his dominant level of thinking.

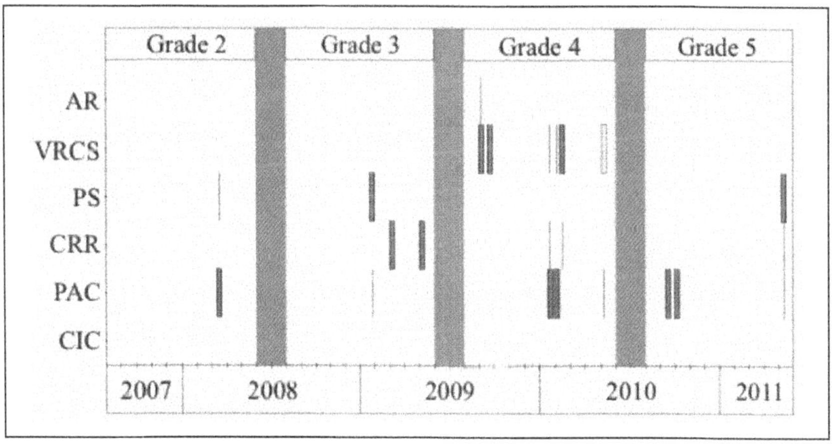

Figure 9.14. Drew's longitudinal growth chart.
CIC = Capacity Indirect Comparer, PAC = Primitive 3-D Array Counter, CRR = Capacity Relater and Repeater, PS = Partial 3-D Structurer, VRCS = 3-D Row and Column Structurer, AR = 3-D Array Structurer.

DISCUSSION AND SUMMARY

In our inspection of the composite growth chart (see Figure 9.15) and the individual charts for children's reasoning along the volume trajectory, as well as from our retrospective review of the teaching-experiment data for the children at the Midwest site, we observed four themes. First, we saw that thinking at the 3-D Row and Column Structurer level was the dominant level of thinking for all five children during Grade 4. For three of the five children, it was the only dominant level observed. Furthermore, three of the five children exhibited thinking at the 3-D Array Structurer level during Grade 5.

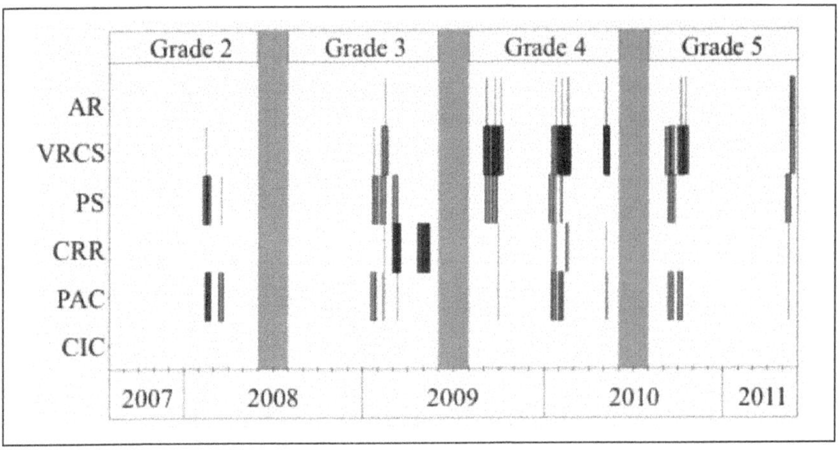

Figure 9.15. Composite growth chart. CIC = Capacity Indirect Comparer, PAC = Primitive 3-D Array Counter, CRR = Capacity Relater and Repeater, PS = Partial 3-D Structurer, VRCS = 3-D Row and Column Structurer, AR = 3-D Array Structurer.

Second, the Capacity Relater and Repeater level appeared as a dominant level of thinking for all the children at the Midwest site during the second half of Grade 3 but not after Grade 3 (at least not as a dominant level). In contrast, the Northeast cohort of children exhibited the Capacity Relater and Repeater level by Grade 2 (see Figure 8.11 in Chapter 8). The researchers attributed this early onset for the Northeast cohort, at least in part, to the children's opportunity to receive focused measurement instruction from pre-K to Grade 2 because of their participation in the study.

Third, some of our data strongly supported the gradual development of sophistication in thinking about volume measurement, as shown by the growth charts for two of the five cohort children, Abby and Arielle. The overall trend was less pronounced for the remaining three children; however, they also showed some upward progress and a range of levels. In particular, two of the remaining three children exhibited a shift in thinking from the Partial 3-D Structurer level to the 3-D Row and Column Structurer level over the 4 years. One child, Drew, had a more complicated overall trend because of an interaction among the components of the area and volume LTs.

Finally, all five children in the Midwest cohort experienced at least some fallbacks during this 4-year study. Most of these fallbacks occurred in response to individual tasks and did not constitute a shift in dominant thinking level. We hypothesize two possible explanations for these fallbacks—explanations that may both have been in play because they are complementary. On one hand, the volume-measurement tasks may have overchallenged or underchallenged children. These discrepancies in the level of the challenge may have been attributable to the way that the researcher presented or represented the three-dimensional figure, or they may be attributable to the task context. In either case, the result may have been that a child used a less sophisticated level of thinking to solve a task. On the other hand and in keeping with our theoretical perspective of hierarchic interactionalism (Sarama & Clements, 2009b), the dominant level may be tentative at first for a child. Eventually, this new level gains ascendancy as the most common way of thinking for that child. In Chapter 10, we present a comprehensive revision of the LT for volume that we introduced in Chapter 7, combining findings reported here with those described in Chapter 8 in a summary of our present understanding of children's development of volume measurement.

Chapter 10

Integration of Results: A Revised Learning Trajectory for Volume Measurement

Douglas W. Van Dine
Metropolitan State University of Denver
Douglas H. Clements
University of Denver
Jeffrey E. Barrett
Illinois State University
Julie Sarama
University of Denver
Craig J. Cullen
Illinois State University
Melike Kara
Towson University

The quantitative results from the Rasch modeling reported in Chapter 7, as well as the qualitative results from empirical evidence reported in Chapters 8 and 9, indicate that the initial developmental progression of the learning trajectory (LT) for volume measurement was generally an effective tool for predicting and analyzing children's behaviors and mental actions on objects, as well as for identifying shifts in children's thinking and behaviors for volume measurement. The results also indicated the need for refinement and revision of this developmental progression, including subtrajectories for different task structures. In this chapter, we review the need for expansion of the initial developmental progression and introduce a revised developmental progression of the LT for volume measurement.

Extensions: Volume Subtrajectories

We observed children in our study, especially the children in the Northeast cohort, exhibiting behaviors related to what appeared to be distinguishable subtrajectories of volume filling, packing, building, and comparing. For example, we observed that some children recognized that filling objects was possible (i.e., capacity recognition), although they did not necessarily recognize that objects take up space (i.e., volume recognition). On the basis of our work in the present study, we assert that number sense connects with (and eventually integrates with) these volume subtrajectories; during early teaching episodes, children did not appear to connect counts of discrete objects with volume (i.e., the space occupied by an object). For example, when these children compared objects by using units of different sizes, they often showed no dissonance about the size of the unit when determining the volume and simply used the number of units counted to compare the volumes of the objects; they were confident that the container with more cubes was bigger, regardless of the size of the cubes. Further analysis of children's behaviors revealed that filling and packing, like other subtrajectories, did not develop in rigid order but instead developed in parallel, loosely coupled.

Patterns of thinking on comparing tasks also did not fit into separate levels, as we had hypothesized (i.e., Capacity Direct Comparer and Capacity Indirect Comparer). We saw that as children's understanding of volume became more sophisticated, their comparisons of volume also became more sophisticated, moving from simple direct comparison to comparison of units needed to fill the amount of space within each container. We identified a single developmental progression; but at each level, we saw children demonstrate behaviors related to volume filling, packing, building, and comparing. Therefore, we revised the early levels of the developmental progression to incorporate volume filling, packing, building, and comparing subtrajectories within a single coherent developmental progression of an LT for volume measurement.

Although we saw a need to describe subtrajectories related to volume filling, packing, building, and comparing in the early levels of the trajectory, data from the children in the Midwest cohort did not support such differentiation beyond the Initial Composite 3-D Structurer level. Instead, beginning at the 3-D Row and Column Structurer level, children integrated what had been four separate volume subtrajectories into a single trajectory for volume (consistent with hierarchical interactionalism; Sarama & Clements, 2009). For example, they could consider discrete objects, such as unit cubes, as the unit of measuring space in filling or comparing volume. Therefore, we did not separate revisions at the 3-D Row and Column Structurer level and beyond into subtrajectories related to volume filling, packing, building, and comparing.

Connections Across the Trajectories

We also noticed inconsistencies between the developmental progressions for length and area measurement when we compared them with the developmental progression for volume. For example, in the developmental progression for length measurement, children demonstrated end-to-end length measurement consistently; but although a similar behavior occurred with volume (i.e., children completely packed containers with cubes), the initial developmental progression of the LT for volume measurement did not represent this behavior. When a child develops an understanding of spatial measurement (whether it be length, area, or volume) a similar developmental progression is evident, progressing through recognizing the attribute; comparing (both directly and indirectly); filling the space; quantifying; recognizing the unit; and building, maintaining, and manipulating a unit of units (or a unit of units of units). We sought to identify consistent behaviors and mental actions on objects across all three trajectories to ensure that we represented all of them in the revised developmental progressions.

HYPOTHESIZING A NEW LEARNING TRAJECTORY FOR VOLUME MEASUREMENT

In the following sections, we examine each level of the developmental progression, incorporating the data from the previous three chapters as well as additional analyses, to present a refined and revised LT for volume measurement. Our focus at each level is on both confirmation and modification.

Revisiting Volume Quantity Recognizer

Our evidence supported the claim that children begin their progression toward a complete understanding of volume by recognizing volume as a quantity, characterized by their mental actions on objects of perceiving space and knowing that objects occupy space. All the children in our study, including the children in pre-K at the Northeast site, demonstrated thinking at and above the Volume Quantity Recognizer level on the initial assessment. Even at this initial level of volume recognition, we identified four subtrajectories related to volume: filling, packing, building, and comparing. In both filling and packing contexts, children made statements that indicated that they perceived space inside objects, such as "This glass holds a lot of water" or "This box holds a lot of toys," but did not report a quantity (see the discussion of Lia's response to the lemonade task in the Volume Quantity Recognizer section in Chapter 8). In describing or building objects, children used general vocabulary such as *big* or *small* to describe overall size; and when the researcher gave a child an object constructed of cubes, the child often counted only what she or he saw on one face of the object. Comparisons of volume were gross comparisons, if children made them, or children focused on linear extent and identified an object that was larger in one dimension (often height) as having a larger volume.

Replacing Capacity Direct Comparer and Capacity Indirect Comparer With Volume Filler

As previously stated, comparison of volume does not happen at one distinct level. Instead, as children develop a more sophisticated understanding of volume, the ways in which they compare the volumes of objects become more sophisticated. Children may begin by making gross comparisons of volume, then move to comparing directly in only one dimension, and eventually compare by interrelating quantity involving three dimensions. Because comparison of capacity is only one type of volume comparison, including individual levels that focus specifically on capacity is not consistent with our analysis of children's behaviors. In the initial developmental progression, Sarama and Clements (2009b) posited that the one-dimensional nature of filling implied the existence of a separate level that developed early. However, evidence from this study with children did not support the existence of separate prerequisite levels.

Instead of individual levels, we saw the need to incorporate the early development of comparison, as well as the movement toward more sophisticated comparisons, into the levels of the revised LT for volume measurement. In addition, we have included both the comparison of volume by using spatial structuring and the comparison of capacity. We therefore concluded that Capacity Direct Comparer and Capacity Indirect Comparer were not independent levels within the revised LT for volume measurement but that we should instead incorporate them into the Volume Filler level (see Table 10.1).

Table 10.1
A Comparison of the Levels From the Initial Progression and the Revised Developmental Progression for Volume Measurement

Initial Developmental Progression		Revised Developmental Progression
Volume Quantity Recognizer	→	Volume Quantity Recognizer
Capacity Direct Comparer Capacity Indirect Comparer	} →	Volume Filler
Primitive 3-D Array Counter	→	Volume Quantifier
Capacity Relater and Repeater	→	Volume Unit Relater and Repeater
Partial 3-D Structurer	→	Initial Composite 3-D Structurer
3-D Row and Column Structurer	→	3-D Row and Column Structurer
3-D Array Structurer	→	3-D Array Structurer

As a further modification, we observed behaviors in children that were representative of mental actions more sophisticated than those in the Volume Quantity Recognizer level and less sophisticated than those in the Primitive 3-D Array Counter level. We therefore modified the LT for volume measurement by placing the Volume Filler level between the Volume Quantity Recognizer level and the Primitive 3-D Array Counter level. At the Volume Filler level, children can directly compare both capacities and volumes by using multiple dimensions but may not explicitly recognize and integrate all three dimensions of an object or container. This recognition of more than one dimension may be also evident in children's counting to determine the volume of a three-dimensional object constructed of cubes. At this level, we saw that children recognized and counted cubes on more than one face, but they still did not count cubes exhaustively on all faces.

The Volume Filler level meets another of our goals; that is, the leveled account should emphasize common aspects of development within related content topics (as is true here for measures of length, area, and volume). We see the Volume Filler level as parallel in its characterization of children's actions to repeat small portions of a measured object (physical unit pieces) until the object to measure is completely exhausted or filled (with units). The typical actions that we observed for the Volume Filler level were parallel to children's actions described within the End-to-End Length Measurer level for the domain of length and the Physical Coverer and Counter level (previously Side-to-Side Area Measurer) for the domain of area. Children at this level can visualize and expect that filling three-dimensional space with objects is possible; further, children at this level typically exhibit an ability to build, maintain, and manipulate a mental image of an amount of material that is space filling when they have perceptual support. We therefore incorporated the idea that children can fill a container and recognize the container as being filled into the Volume Filler level.

Redefining Primitive 3-D Array Counter as Volume Quantifier

Primitive 3-D Array Counter is the level at which spatial structuring begins to be evident; and children develop an explicit, but partial, understanding of how they can arrange cubes to fill space. As described in Chapters 7 and 8, we found confirmation of this level of thinking occurring after Volume Filler because children in the Northeast cohort exhibited this level as a dominant level toward the end of kindergarten and into Grade 1. All the children in the Midwest cohort were at or above this level at the beginning of the study when in Grade 2. Children counted the faces of a cube building, often double-counting at the edges and vertices, but they did not demonstrate recognition of hidden or internal cubes in their counting (see the descriptions of Zola, Robert, and Lia comparing the four blocks in the Primitive 3-D Array Counter section in Chapter 8 and the description of Abby comparing the solids depicted in Figure 9.1 in Chapter 9). With guidance and instruction, however, the children were able to count cubes one at a time. If the researchers gave the children prisms with grids marked on the faces as indicators of cube arrangements and the prisms had no hidden interior cubes, the children counted the cubes correctly.

As with previous levels, however, packing and building constitute different contexts for young children, and this level addresses each context. During packing tasks, children were able to pack a box completely and neatly with cubes, but they were only able to determine a total when counting occurred while they were packing. If a researcher asked children to determine the volume after the packing was complete, they might not account for cubes that they were unable to see. Children at this level of thinking moved to account for all faces of a three-dimensional object when building, whereas they had previously shown only partial recognition of multiple faces.

In comparison contexts, children at this level used direct comparisons but recognized that a three-dimensional structure has three dimensions. They also attempted to use the number of cubes as a way to compare the volumes of two three-dimensional structures constructed of cubes. However, the children's comparisons focused on the number of cubes in each structure, and they inconsistently recognized the importance of the size of the cubes. That is, children compared volume on the basis of the number of cubes that they counted in each structure and did not consider that a larger unit cube would result in a larger volume, if the same number of cubes were in a structure (see Figure 10.1). In filling tasks, many of which involved cylindrical containers, children continued to refer to the height as one dimension but moved beyond simply referring to the width as an additional dimension. Some children recognized the circumference by placing their fingers into a circular shape, whereas others appeared to refer to the cross-sectional area by placing a hand inside the top of the container and saying, "It's bigger here."

Overall, we found solid evidence to confirm the Primitive 3-D Array Counter level. We did, however, determine that the name *Volume Quantifier* was more appropriate to describe this level of thinking. Children at this level focused on quantity (i.e., the number of cubes) as volume and were not making an explicit connection between the number of cubes and the space filled.

Figure 10.1. Volume comparison with units of different sizes.

Redefining Capacity Relater and Repeater as Volume Unit Relater and Repeater

All the children in the Midwest cohort exhibited thinking at the Capacity Relater and Repeater level beginning in Grade 3, whereas the children in the Northeast cohort progressed to thinking at the Capacity Relater and Repeater level during Grade 1. For example, children in the Northeast cohort exhibited this level of thinking when they filled a container by repeatedly filling a unit container and counting how many units they needed. These children also recognized that they would need fewer large objects or units than small objects or units to fill a given container (see the discussion of Robert and Lia filling a $2 \times 2 \times 3$ container with a $1 \times 1 \times 1$ cube scoop in the Capacity Relater and Repeater section in Chapter 8). In addition, in filling tasks, we saw that children related the number of units needed to fill a container to the height of the container; for example, when determining how many glasses of water would be necessary to fill a container, some of the children iterated a height interval for the number of glasses filled after they had poured in one glass. We extended the action scheme of this level to include relating the number of units to the height of the container to fill with that unit.

For example, we presented children with two rectangular prisms, one constructed by using wooden 1-inch cubes (two cubes by four cubes by two cubes) and the other of yellow 1-centimeter cubes (two cubes by three cubes by four cubes), along with the following situation: "Another student, when asked to compare the volume [of these two blocks], said that the yellow block has a greater (bigger) volume because it is made up of more cubes. Do you agree with this student?" In this case, the prism made of centimeter cubes was obviously smaller in volume (see Figure 10.1). The results indicated that children who attended to the unit size in relation to the expected overall measure were able to explain the role of the size of the unit in their comparison of volumes as an inverse relationship. Because of this evidence as well as similar evidence, we expanded the level to include all four subtrajectories related to volume measurement (i.e., filling, packing, building, and comparing).

Redefining Partial 3-D Structurer as Initial Composite 3-D Structurer

While the spatial structuring abilities of children continued to develop, we saw their thinking move to align with that of the Partial 3-D Structurer level. The children at the Northeast site demonstrated thinking at that level in Grade 2 toward the end of the study, whereas the children at the Midwest site did so sometime between Grade 2 and the beginning of Grade 4. At the Partial 3-D Structurer level, children began to recognize hidden or internal cubes, and their cube counting became more accurate in reporting volume. At first, they attempted to account for individual hidden cubes when counting. Over time, children exhibiting thinking at this level came to recognize the number of cubes in either a row or a column (what we call a *core* or $1 \times 1 \times n$ *composite unit*) and used additive reasoning or skip counting to determine the total. Children therefore moved from recognizing individual units to building, maintaining, and manipulating mental images of composite units. Our evidence confirmed that this recognition was evident in both packing tasks (see the discussion of Anselm's solution to the box task displayed in Figure 9.5 in Chapter 9) and building tasks (see the discussion of how Edith quantified the volume of the $4 \times 3 \times 3$ rectangular prism in the Partial 3-D Structurer section in Chapter 8 and the discussion of how Abby built a $3 \times 2 \times 2$ rectangular prism in the Partial 3-D Structurer section in Chapter 9). Hence, we decided to rename the level *Initial Composite 3-D Structurer* because it was a more accurate description of this shift in thinking and its placement within the LT.

These children continued to develop in their thinking regarding filling volume as well as in their ability to compare volumes. Over time, children moved beyond characterizing volume as a count of cubes and came to understand volume as space. For example, in filling and packing contexts, they began to focus both on the amount filled and the space remaining to be filled, thereby developing a sense of when a container was half full (see the discussion of Abby working with the prisms shown in Figure 9.3 in Chapter 9). In addition, they began making connections between filling volume and packing volume and understood that the cubes actually take up space, as liquid or fluid material does. As an example, during Grade 2 at the Northeast site, we presented children with tasks involving measuring containers marked in units equivalent to the blocks with which they were packing (i.e., if packing with inch cubes, the measuring container was marked in cubic inches). After establishing the equivalency with children, we asked them to fill one container with rice and then use the measuring container (marked in cubic inches) to determine its volume by pouring the rice from the filled container into the measuring container. We also asked the children to measure the volume of another container by packing it with inch cubes. In these tasks, children demonstrated an ability to relate the number of cubes to the amount of space filled.

Revisiting 3-D Row and Column Structurer

All the children in the Midwest cohort exhibited thinking at the 3-D Row and Column Structurer level during the study. In contrast, only two children in the Northeast cohort exhibited that level of thinking—and then only at the end of Grade 2. We found evidence that children at the 3-D Row and Column Structurer level used counting strategies in which they visualized rows and columns, as well as layers of cubic units. They also used additive reasoning or skip counting to determine the number of cubes in an array (see the discussion of Anselm completing the task of filling a box when given one layer and one core, as presented in Figure 9.7). At times, children at the 3-D Row and Column Structurer level used multiplication instead of skip counting or addition. Our evidence showed that children counted the number of cubes in one layer by counting individual cubes, but they still used multiplication to find the volume by multiplying the number of layers times the number of cubes from each layer (see the discussion of Abby completing the gridded packing task presented in Figure 9.6). We also saw that children used multiplication (i.e., the number of cubes in a row times the number of rows in a layer) to determine the number of cubes in one layer but then used repeated addition, skip counting, or doubling strategies to determine the total number of cubes in the prism.

We found evidence of children's thinking about the decomposition of arrays and have used this evidence to refine the level description. Children operating at this level can decompose layers into cores (i.e., rows or columns) and decompose cores into cubes. They struggled, however, when composing arrays from small collections of units: Finding the number of smaller arrays of cubic units needed to fill a larger array was still difficult for them. We also saw that children operating at this level may show a propensity for additive reasoning in volume comparisons. When we carefully selected the wording of the comparison task, we found that children were capable of engaging in comparisons as a ratio (i.e., a multiplicative, rather than an additive, comparison; see Cullen, Miller, Barrett, Clements, & Sarama, 2011).

Observations of children's misconceptions provided extensions of the initial developmental progression because children operating at this level often did not have parallel subtrajectories related to different task representations. For example, we found that although children were able to resolve three-dimensional forms of the tasks, they sometimes added the lengths of the three dimensions instead of using spatial structuring schemes when we posed tasks involving two-dimensional representations of three-dimensional figures. We also observed that children often made iteration or counting mistakes when decomposing three-dimensional arrays to subcollections of units such as short columns of three or four units taken as a package that must not be broken. This kind of packing task proved difficult for children at this level.

Revisiting 3-D Array Structurer

Although all the participants at the Midwest site demonstrated thinking at the 3-D Array Structurer level, none of the children at the Northeast site did. Our evidence confirmed that children at that level use action schemes for multiplication to determine a measure for volume, as stated in the initial developmental progression. In addition, we saw a need to expand the description of this level to include how children at this level dealt with units of units or units of units of units of volume. For example, children at the 3-D Array Structurer level operated flexibly with horizontal or vertical layers while determining the total number of cubes in an array and used multiplication for both situations (see the discussion of Abby's reasoning in the 3-D Array Structurer section in Chapter 9). They also decomposed three-dimensional arrays into smaller three-dimensional arrays that were not rows, columns, or layers; and they could calculate the number of smaller arrays that would fit into a larger array. Moreover, children operating at the 3-D Array Structurer level were able to flexibly coordinate their multiplicative and additive reasoning for volume comparisons.

REVISED DEVELOPMENTAL PROGRESSION

We confirmed many components in our initial developmental progression (Clements & Sarama, 2009; Sarama & Clements, 2009b) through this study, such as the general progression and the importance of spatial structuring. Other components from the other developmental progressions for length and area measurement played a catalytic role in the addition or refinement of several levels of the revised developmental progression. Validation of these additions and refinements, as well as other refinements and changes, stemmed directly from this theoretical and empirical work. For example, we identified different subtrajectories of understanding volume, and these subtrajectories helped us integrate the previously separated filling (capacity) and packing and building (volume) levels of the initial developmental progression of the LT for volume measurement.

Figure 10.2 shows the developmental progression, mental actions on objects, psychological foundations, and anticipated partial conceptions for a revised developmental progression of the LT for volume measurement as a result of our data and building on the initial LT that we constructed as a synthesis of previous research in children's understanding of volume measurement.

Developmental Progression	Mental Actions on Objects	Anticipated Misconceptions or Partial Conceptions
Volume Quantity Recognizer (VQR)		
Filling and Packing. Recognizes capacity as an attribute, beyond implicit acting on materials. *"I can pour lots of sand into this can."* *"This box holds a lot of toys."* *Building.* Builds with blocks, associating more blocks with terms like *big* and fewer blocks with terms like *small.* *"I used a lot of blocks and made a big house. See?"* *Comparing.* Initially, recognizes volume as an attribute, describes objects with words such as *big, small,* and *tiny.* Eventually, may compare volume, recognizing only one dimension. *"This one's bigger, with more blocks. See, it's taller."* [Comparing two buildings with the same cross section]	Perceives three-dimensional space and objects within the space.	When building, may not accurately recreate a given shape in size (e.g., number of blocks) or dimensions. When given a three-dimensional object constructed of cubes, may count only one face. May not compare, then begins to compare capacities but makes no reference to dimensions.

Figure 10.2 continues on next page

Developmental Progression	Mental Actions on Objects	Anticipated Misconceptions or Partial Conceptions
Volume Filler (VF)		
Filling. Fills a container by using another (smaller) container and counts the number needed to completely fill the larger container. *Packing.* Places cubes into a rectangular box to fill it; eventually packs entire box, with a focus on leaving no gaps. *Building.* Given a three-dimensional object constructed of cubes, recognizes and counts cubes (the child may be counting "blocks" or even "squares") on multiple faces. *Comparing.* Compares objects by physically or mentally aligning them; refers to at least two dimensions of objects. *Places two objects next to each other, saying, "This one holds more because it's taller and it's wider."* Begins to associate number of scoops or cubes in comparison, simply using a bigger number for larger container.	With perceptual support, can visualize that filling three-dimensional space with objects (e.g., cubes) is possible. Can build, maintain, and manipulate a mental image of a particular amount of material ("stuff"). With the immediate perceptual support of the containers and material, can compare such images at an intuitive level. Uses perceptual objects and internal innate competencies to compare extent of three-dimensional space; usually focuses attention on only one or two dimensions. Recognizes overflow as indicating that the container from which the child is pouring contained more than the container into which the child is pouring. May apply explicit transitive reasoning in volume-comparison situations.	May count when filling or packing, but does not focus on quantifying the total volume or capacity. When filling, counts each "scoop" as one, although may not completely fill (or may overfill) the scoops each time. Attends to space filled without comparing it to the total capacity of the container. For example, when filling or packing, may not recognize container as half full. When packing, focuses on filling the container rather than on spatial structuring. May use the walls of the container to constrain the covering of the bottom but may believe that the container is full when only the bottom is covered. Given a three-dimensional object constructed of cubes, may not count on all faces consistently or follow a pattern, may double-count, and may ignore internal cubes. When attempting to reproduce rectangular prisms, only considers some characteristics of shape (e.g., one or two dimensions); therefore, may not produce identical shapes. May not recognize two shapes as having equal volume when researcher orients or constructs them differently.

Figure 10.2 continues on next page

Developmental Progression	Mental Actions on Objects	Anticipated Misconceptions or Partial Conceptions
Volume Quantifier (VQ) (previously Primitive 3-D Array Counter): (VQ) *Filling.* Estimates number of scoops needed to fill. Attends to both the portion of container filled and the portion remaining unfilled. Recognizes when container is half full. *Packing.* Exhibits initial spatial structuring. Packs box neatly and completely with cubes; counts one cube at a time while packing to determine total. *Building.* Exhibits initial understanding of cubes as filling a space (completely and without gaps). Counts on all faces of a three-dimensional object constructed of cubes, has a developing sense of the cube as a unit, and begins to recognize that squares on adjacent faces of a rectangular prism that share a side are faces of the same cube. *Comparing.* Compares objects by physically or mentally aligning and explicitly recognizing three dimensions (Note: Although cylindrical containers can vary in only two dimensions, height and radius, children at this level demonstrate understanding of three-dimensional space.) "*This cup*" [cylindrical container] "holds *more because it is taller and it is bigger here*" [putting hand into the opening at the top to indicate the size of the opening]. Compares the volume of objects by counting the number of cubes, showing initial understanding of cubes as filling space (as described previously); may break a larger object into smaller pieces in order to "see" all the cubes. Recognizes that objects can look different but still contain the same number of cubes.	With strong guidance and perceptual support from prestructured materials, can direct the filling of three-dimensional space with objects (e.g., cubes) and recognize that filling as complete. Implicit visual patterning and constraints of physical materials guide placement of cubes. Begins to associate multiple exposed faces of a single cube ("block") with a single cubic unit. Builds, maintains, and manipulates mental images of amount of material ("stuff") such as liquid. With the immediate perceptual support of the containers and the material, action schemes allow comparisons of (gross) images. *Notes on psychological foundations: Not yet able to understand the basis for the multiplicative transformation of lengths into volume measurements.*	May not recognize the need for units of equal size when measuring. May attempt to iterate a given unit throughout a three-dimensional space but not maintain equal unit size or spacing. May not be able to accurately visualize and calculate total, possibly double-counts cubes at the corners and often does not count internal cubes. May not make an explicit connection between number and space (volume as number).

Figure 10.2 continues on next page

Developmental Progression	Mental Actions on Objects	Anticipated Misconceptions or Partial Conceptions
Volume Unit Relater and Repeater (VURR) Relates size and number of units explicitly to measure; understands that will need fewer larger units than smaller units to fill or pack a given container. Accurately converts units in 1:2 ratio. *Filling.* Uses simple units to fill containers with accurate counting, completely filling the scoop each time. After pouring one unit into the container, anticipates the volume of the container by iterating the height filled by the unit exhaustively along the height of the container. *Packing.* Uses discrete units to pack a container without gaps and with accurate counting. Iterates unit throughout volume, maintaining equal unit size and spacing. *Building.* Exhibits developing understanding of cubes as filling a space. Counts cubes, not faces (or "faces-as-cubes"). *Comparing.* Describes correctly the relative volumes of objects by reasoning about unit size when comparing two three-dimensional objects in situations such as congruent objects containing different numbers of units, noncongruent objects containing the same number of units, or noncongruent objects containing different numbers of units.	Action schemes include the ability to iterate a mental unit through a perceptually available object. Can maintain the image of each placement while moving the physical unit to the next iterative position (initially with weaker constraints on this placement). Stronger constraints on object counting (counts all objects once and only once), and use of rows as an intuitive structure or explicit application of labeling as marker allows child to keep track. These action schemes allow the application of counting-all addition schemes to apply to measures. *Notes on psychological foundations: This level is foundational to the later development of the multiplicative transformation of lengths into volume measurement. This type of multiplicative structure is analogous to n times 1 cubic unit (e.g., to think of 12 cubic units to be placed and counted one at a time as 12 times 1 cubic unit).*	May overgeneralize unit iteration of length to containers with nonuniform cross section, i.e., focusing only on the one dimension, such as height and not on variations in cross-sectional area. May not accurately convert units in a ratio other than 1:2. In packing or building, may not account for internal or hidden cubes. In comparing tasks, may not explicitly quantify discrepancies.

Figure 10.2 continues on next page

Developmental Progression	Mental Actions on Objects	Anticipated Misconceptions or Partial Conceptions
Initial Composite 3-D Structurer (VICS) Understands cubes as filling a space. Explicitly relates size and number of units to volume. Uses additive reasoning (e.g., skip counting to obtain total). Conversion of units develops to include ratios other than 1:2. *Filling.* Relates number of cubes to cubic units as measured by capacity. *Given a graduated cylinder marked in cubic-inch units, child understands that sand filled to the 10 in the cylinder would fill a box that holds ten 1-inch cubes.* *Packing.* Begins to visualize and operate on composite units such as rows or columns (what we call a $1 \times 1 \times n$ *core*). Iterates to pack the space completely, accounting for internal or hidden cubes. Decomposes space, allowing for accurate use of units and subunits. Recognizes when a box is half full; visualizes remaining rows or columns. *Building.* Develops more accurate counting strategies. Counts systematically, accounting for internal or hidden cubes, and moves to operating on composites, including rows and columns. *Comparing.* Connects volume as number and volume as space. Develops sense of conservation for the cases that involve transformation.	Builds, maintains, and manipulates mental images of composite three-dimensional shapes, structuring them as combinations of individual shapes and as a single entity—a row or column (a unit of units). Applies this composite unit repeatedly, but not necessarily exhaustively, because intuition continues to guide its application. *Notes on psychological foundations: This level anticipates later development of the multiplicative transformation of lengths into volume measurement. This type of multiplicative structure can be analogous to n rows times m cubic units per row (e.g., for 12 cubic units built, packed, or counted, in part, as three rows with four cubic units per row).*	May not recognize the disparities between the measure of length of an edge and the number of volume units fitting along that edge. Focuses on either a row or a column but cannot transition from one to another. Does not visualize and operate on a layer as a composite.

Figure 10.2 continues on next page

Developmental Progression	Mental Actions on Objects	Anticipated Misconceptions or Partial Conceptions
3-D Row and Column Structurer (VRCS) Coordinates flexibly filling, packing, and building schemes of volume. Shows a propensity for additive comparisons (e.g., "this one has 12 more") but may show some nascent multiplicative comparisons (e.g., "this one is four times as big"). Initially counts or computes (e.g., number of rows times number of columns) the number of cubes in one layer ($1 \times m \times n$) and then uses addition or skip counting by layers to determine the total volume. Eventually moves to multiplication (e.g., number of cubes in a layer times number of layers). Operates fluidly and flexibly on units (cubes), units of units (rows or columns), and units of units of units (layers). Composes and decomposes array ↔ layers ↔ rows/columns ↔ units. With perceptual support, can decompose three-dimensional arrays into other complex three-dimensional arrays (not only layers, rows, or columns) and calculate the number of these smaller arrays in the larger array.	Builds, maintains, and manipulates mental images of composite units, operates on them as composites of individual shapes and as a single entity—a layer (a unit of units of units) of congruent cubes. Applies this composite unit repeatedly and exhaustively to fill the three-dimensional array, coordinating this movement in 1:1 correspondence with the elements of the orthogonal column. May apply a skip-counting scheme to determine the volume. In a measurement context, applies the concept that the length of an edge specifies the number of volume units that will fit along that edge but may need to create a perceptual layer to support reasoning. *Notes on psychological foundations: This level is the foundation for the multiplicative transformation of lengths into volume measurement. This type of multiplicative structure can be analogous to n layers times m cubic units per layer.*	When decomposing three-dimensional arrays into complex arrays, may still make iterating or counting mistakes, especially if the decomposition case requires mental transformations (e.g., rotations). Although visualizes and operates on horizontal or vertical layers, may not operate flexibly on both. When determining volume, may add the edge lengths rather than apply an appropriate mental structuring.

Figure 10.2 continues on next page

Developmental Progression	Mental Actions on Objects	Anticipated Misconceptions or Partial Conceptions
3-D Array Structurer (3-D AS) Has an abstract understanding of the volume formula for a rectangular prism. Shows a propensity for multiplicative comparisons, coordinates multiplicative and additive comparisons flexibly. With linear measures or other similar indications of the three dimensions, multiplicatively iterates cubes in a row, column, or layer to determine volume. In multiple contexts, computes the volume of rectangular prisms from their dimensions and explains how that multiplication creates a measure of volume. Visualizes and operates on both horizontal and vertical layers, even without perceptual support. Decomposes three-dimensional arrays into other complex three-dimensional arrays (not only layers, rows, or columns) and calculates the number of these smaller arrays in the larger array (by using repeated addition, multiplication, or division). Coordinates the spatial and symbolic decompositions.	Builds, maintains, and manipulates composites (a three-dimensional array—units of units of units of units) and operates on three dimensions. Mentally composes and decomposes a) array⟷layers⟷rows/columns⟷units b) array⟷smaller complex arrays⟷units. Can visualize and operate on arrays spatially or symbolically. Can connect $(l \times w) \times h$ with number of cubes in a horizontal layer times number of horizontal layers and $l \times (w \times h)$ to number of cubes in a vertical layer times number of vertical layers. Curtails the process to use volume formulas with understanding. In a measurement context, applies the concept that the length of an edge specifies the number of volume units that will fit along that edge, even without perceptual support. *Notes on psychological foundations: The previous levels were foundational for the development of multiplicative transformation of lengths into volume measurement, which have been developed in tandem.*	

Figure 10.2. Revised developmental progression component of the LT for volume measurement.

SECTION IV

Chapter 11

Reflections on the Design and Outcomes of the Research

Jeffrey E. Barrett
Illinois State University
Julie Sarama
University of Denver
Douglas H. Clements
University of Denver
Craig J. Cullen
Illinois State University
Douglas W. Van Dine
Metropolitan State University of Denver

In the preceding chapters, we described our study of the growth of children's thinking and learning about geometric measurement from pre-K to Grade 5. Extant research (e.g., Battista & Clements, 1996; Curry & Outhred, 2005; Hiebert, 1981; Outhred & Mitchelmore, 2000; Piaget, Inhelder, & Szeinska, 1960; Stephan & Clements, 2003) guided the development of initial learning trajectories (LTs) for length, area, and volume measurement (Clements & Sarama, 2009; Sarama & Clements, 2009b). We used the developmental progressions of these initial LTs to compile longitudinal accounts of children's increasingly sophisticated thinking about length, area, and volume. These accounts gave us evidence on which we based our revisions to the developmental progressions of initial LTs (Clements & Sarama, 2009; Sarama & Clements, 2009b). Our work follows recommendations that researchers need to check progressions and trajectories through longitudinal accounts of children across periods that fit the extent of the trajectory levels (Daro, Mosher, & Corcoran, 2011; C. L. Smith, Wiser, Anderson, & Krajcik, 2006). In this chapter, we reflect on the design and implementation of our study, emphasizing the challenges and opportunities that we encountered by studying developmental progressions that encompass the 7 years of development from pre-K to Grade 5.

We begin this chapter with a description of the processes of designing tasks and analyzing children's thinking and behaviors in the spatial measurement domain by using LTs (Clements & Sarama, 2014b; Cobb & Gravemeijer, 2008; Simon, 1995, 2006; Simon & Tzur, 2004; Sztajn, Confrey, Wilson, & Edgington, 2012). We describe the dual role of a learning trajectory (LT) in our research: We used LTs as theoretical tools that guided instruction, assessment, and analysis while we worked to improve those same LTs (Cobb, Confrey, diSessa, Lehrer, & Schauble, 2003; Confrey & LaChance, 2000; Ruthven, Laborde, Leach, & Tiberghien, 2009).

We then describe the challenges associated with integrating multiple methodologies (qualitative analyses of teaching episodes to examine quantitative analyses of cross-sectional assessment data) across multiple sites during a 4-year longitudinal study. To do so, we discuss the following questions: How can we build tasks that simultaneously test LTs and help children gain knowledge? What are the advantages and disadvantages of various contexts for a teaching experiment? In what ways did children's perceptions about the process change the data collected during teaching episodes? What kinds of strategies help us coordinate the analysis of data across sites? Finally, we shift our attention to discuss implications and directions for future research and educational practice.

THE DUAL ROLE OF LEARNING TRAJECTORIES

To conduct a longitudinal study that focuses on the use and refinement of LTs, we made several critical design and analysis decisions throughout the 4 years of data collection. We designed this longitudinal study to complement extant cross-sectional research about children's thinking in the spatial measurement domain (Sarama & Clements, 2009b). Cross-sectional research supports broad developmental accounts for specific domains of knowledge, but longitudinal and experimental research needs to verify such accounts (cf. Lehrer, Jacobson, et al., 1998; C. L. Smith et al., 2006). That is, the longitudinal design allowed us to chart with more certainty the sequence of levels through which individual children pass. We discuss the extent to which the LTs served as effective tools to support research within a teaching-experiment context—planning appropriately engaging tasks for children, predicting their responses to tasks on the basis of their prior performance, and describing children's thinking in relation to their actions and explanations. To explain the revision and extension process in terms of what counts as adequate evidence to modify the LTs, we describe two related uses of LTs. First, we consider LTs as design objects to build theory, because testing and refining the LTs connect intimately with testing and refining the theory that forms the basis of the LT. Second, we consider LTs as analytical tools, because their use allowed us a lens with which to observe children, in addition to suggesting the creation of related analytical tools, so we could understand the children and the LTs better.

Learning Trajectory as Design Object: Building Theory

We produced empirical accounts of children's learning through the course of teaching episodes and used these accounts to confirm or disconfirm our local working hypotheses about specific children. We developed these hypotheses from the initial LTs while we worked from teaching episode to teaching episode with each child or with groups of children. We used a combination of formal assessments and predictions about levels of understanding for each child during successive teaching episodes to check the developmental progressions of the LTs. The teaching-experiment methodology allowed us to respond or adapt to children's thinking during teaching episodes and check for the correspondence between the hypothetical accounts (i.e., developmental progressions of the initial LTs) and the empirical accounts.

When the findings from the teaching experiment and our predictions based on the developmental progression of the LT suggested disparate outcomes, we had to decide whether we needed to adjust the LT. For example, during the spring of Grade 5, Owen explained how he mentally structured a rectangular region into two columns of three rectangles, each consisting of two triangular units, thereby acting on conceptual (mental) units to measure area (for more information on this task, see Kara, Eames, Miller, Cullen, & Barrett, 2011). When we asked him to represent his thinking with figural (drawn) units, he did not conserve the shape of the triangular unit or correctly structure the rectangular region. We followed these teaching episodes with questions and tasks for measuring areas by using triangle-shaped units. On each of these tasks, we found that he was able to describe how he structured to measure a region with conceptual units, but he did not correctly structure the same region with drawn units. Although Owen's response to these tasks did not follow the progression delineated in the initial developmental progression of the LT for area measurement, we left the developmental progression unaltered because we lacked similar evidence from other cohort children. We took that example as a caution against overgeneralizing or rigidifying the developmental sequence that proceeds from motor-spatial operations to figural operations to conceptual operations (cf. Clements, 1999a; Sarama & Clements, 2009a).

Although we expected to check our generalizations about children's ways of developing knowledge, we often encountered variable progress for specific children (see the growth charts for children in the Midwest and Northeast cohorts in Chapters 2, 4, 5, 8, and 9). This variable progress was a challenging aspect of our work; we found that all children exhibited variable progress at any one developmental moment for a given teaching episode. We took this finding as an important aspect of the data for various children who were subjects in this study, a discrepant event that we sought to assimilate in our efforts to find general progressions for children. However, we noticed that different children exhibited different amounts of variability. In retrospect, we realized that this finding is consistent with the theoretical framework of hierarchic interactionalism because it indicates the potential variability for an individual within a learning situation that includes scaffolding by a teacher. Our construct for an LT-level analysis emphasizes the frequent need to characterize more than one level of performance for a child at any moment. More important, however, we note that we should take the extensible quality of each child's thinking and behaviors as a characteristic aspect of the child's entire profile as a learner, assuming a maximal scaffolding of support from another person. The differences among the capacities of children to reach past their established levels of thinking indicate that teaching should account for differing instructional requirements and supports to avoid underreaching or overreaching the level of thinking for a given child or group of children with a similar learning profile (Ginsburg, 2009; Vygotsky, 1978). LTs help define the "zone of proximal development" by allowing the teacher to (a) ensure that the present dominant level of thinking is well established, (b) then present instructional tasks that are only one level beyond that level, and (c) either slowly build that new level or move to even higher levels.

We note the limitations of our study, given the cohorts' children at each of two sites for the teaching experiment cycles of teaching episodes during 4 years. Although we found the progressions and related sets of tasks productive and tractable for these children, we recognize the need to check these LTs with various subsets of the population to heed sociocultural, socioeconomic, curricular, and local variations in children's developmental patterns. Further research is necessary to explore the effects of such variation on children's patterns for learning measurement; but more extensive research across cultures, economic situations and curricular context is needed (cf. Sarama & Clements, 2009b).

Sometimes we found situations in which our predictions did not match what children did or said from one teaching episode to another. As discussed previously, we had to decide whether to change the developmental progression immediately or wait for further indications that we needed to change the progression. Should we expand the developmental progression to include another description of thinking, or should we continue to work to help the child move or grow as we had predicted? For the most part, we took the latter approach; we commonly waited to learn whether another child would exhibit mental operations and thinking patterns that had not been anticipated or documented within the progression or whether the actions by that child would eventually demonstrate thinking more in line with what we had predicted.

For example, we revised the sequence of the developmental progression of the LT for length measurement to take one level, Indirect Length Comparer, out of the essential pathway of the trajectory because our review of qualitative data showed that several children often moved through a sequence of levels of thinking without any evidence of working at that level. The records for one child showed that of 21 teaching episodes, seven of the episodes indicated a skip for the indirect level, seven included the level, and the other seven were inconclusive. The prevalence of such level skipping by the children in the Northeast cohort warranted our tentative revision of the developmental progression of an LT.

We also extended the description of levels to address a broader set of schemes. For example, we extended the LT for area measurement to incorporate accounts of children's thinking on nonrectilinear regions and with nonsquare units of area. We made this change for two reasons: First, we developed tasks to help children recognize implicit advantages of the operative schemes for area measurement when measuring rectangular regions with square units; second, we wanted to describe the children's thinking on these new task situations that centered on nonsquare units and nonrectilinear regions. The intentional task development resulted in an instructional benefit for three of the children at the Midwest site. This change also enabled us to describe the thinking exhibited by these children by using a broader set of schemes; we found multiple children exhibiting patterned sequences of action that the initial developmental progression had not predicted or described.

Similarly, we determined that the initial developmental progression of the LT for volume measurement had a narrow focus, and we consequently expanded it to distinguish among a range of actions or schemes that children demonstrated and that emerged from these data: filling, packing, building, and comparing quantities.

In summary, we negotiated the resultant descriptions of developmental progression levels through the interaction of existing syntheses of relevant literature and continual use in teaching-experiment cycles of instruction and assessment. We used the level claims to predict and check in the teaching-experiment cycles with two cohorts of children. The revised version of the developmental progression levels, the main component of an LT, constitutes a product of design research that has its basis in broader empirical work, guided by specific longitudinal data from two cohorts of children and grounded in our theoretical perspective of hierarchic interactionalism. We believe that our usual cautious approach of looking for multiple indications before making changes proved effective and practical. Our research team did experience tension over whether we would realize the greatest value by maintaining the LT as an intact theoretical tool for extended periods of testing or whether we should revise it on a more regular basis and keep it closer to the immediate findings in the teaching experiments. We more often followed the former course of action, choosing to maintain the LT for extended testing. This limitation of the present work was attributable to our longitudinal emphasis. Further cross-sectional analysis of a larger sample of children or longitudinal work with larger populations is necessary.

The Learning Trajectory as an Analytic Tool: Analyzing Data and Describing Development

One essential focus for this study was to determine whether longitudinal data would support our generalized accounts of children's increasingly sophisticated levels of thinking (i.e., the initial LTs). To summarize our accounts of children's thinking across the two cohorts, we developed condensed records of the exhibited levels of thinking for each child and presented each case in a growth chart. We built these empirical descriptions by characterizing a child's exhibited behaviors and thinking about measuring from data that included records for specific dates, a specific teaching episode, or a collection of teaching episodes, perhaps as much as an entire semester. We treated these descriptions like novelettes and used varying durations of time and data-collection moments to follow each cohort child. We also used multiple cases to build composite accounts of thinking for a given spatial measurement domain (see growth charts at the close of Chapters 2, 4, 5, 8 and 9). These charts suggest the specific interval of time between subsequent levels of thinking across semesters and across the entire range of seven semesters. At the conclusion of our work with these longitudinal studies, we identified 10 to 12 months as a typical time period to expect a child to shift from one dominant level of thinking along the trajectory to the next, considering the contexts of their experiences, including traditional school experiences and our work with the children.

Our work presented tasks, but not curricular units or intentional teaching strategies per se. We next present our rationale for the conclusion that 10 to 12 months represents the typical time period in which we expect a child to shift from one dominant level to the next.

Several challenges ensued while we worked to characterize patterned shifts in thinking across and within the children in the two cohorts. Although we were using each teaching episode as a data point for one child, we usually had at least three different task outcomes to code a session; at times, they might indicate in divergent levels. How should we categorize the knowledge level for that session? We decided during the second year of our project to require the researcher to report a dominant level for each teaching episode in addition to reporting other levels that he or she observed during the episode. We could isolate this dominant level for any one teaching episode, but what does that claim mean? We studied the data—including the researcher's journal notes, video records, and drawings or notes that the child made—from each teaching episode with each child to identify the most prominent level of thinking exhibited during that teaching episode. The claim of a dominant level meant that we would expect to find the child exhibit that same level during a subsequent teaching episode on the same topic or during observation of the child's classroom activity on a measurement task soon afterward, say, within the next couple of months.

Settling on a dominant level of thinking was often challenging to defend empirically because children exhibited wide variation in behaviors on different tasks, even during the same session. For example, we determined that Abby was using dominantly Partial Row Structurer level thinking during late spring of her Grade 4 year for a given teaching episode, with some evidence of Area Row and Column Structurer, which is a more sophisticated level, and some evidence of the Area Unit Relater and Repeater level, which is a less sophisticated level (see Abby's drawing for the triangular unit task illustrated in Figure 6.1 in Chapter 6). Because we had the opportunity to look retrospectively across 4 years and 10 cohort children, we argue that each child tends to exhibit a consistent pattern across several teaching episodes during the same semester. Further, many of the children exhibited the same level of thinking for approximately 10 to 12 months before exhibiting thinking at the next or any other level of the LT (see the growth charts at the close of Chapters 2, 4, 5, 8, and 9 for examples). However, this 10- to 12-month timeframe for each level of the LT was not apparent to us until we made our retrospective analyses. One implication is that researchers should help teachers become aware of progress within the present level of thinking as well as nascent learning from the subsequent level. Another implication is that future research should investigate whether students can make faster progress with better curricula and pedagogy—especially with more attention to—and integration of—measurement experiences in mathematics and across the curriculum.

The retrospective analysis stage involved reviewing collections of observations across multiple teaching episodes to look for developmental shifts rather than short-term or incidental learning shifts. Further, to generalize our observations about development, we looked across several cohort children to examine broad developmental shifts. We therefore coordinated multiple growth charts along the same axes to represent the collective progress of the children in both cohorts. Most broadly, we summarized the patterns of thinking across different children to compare our findings with the LTs, in keeping with design research (Cobb et al., 2003). The individual growth charts and the composite growth charts summarize patterns of change among the children in this longitudinal study, justifying warrants for refinements to the LTs. This pattern of gradual incremental development and the identification of a dominant level bracketed by a range of fallback levels and advances present during any one teaching episode is consistent with our theoretical framework of hierarchic interactionalism.

On the basis of our analytical experiences through this research project, we recommend that a researcher who is conducting a teaching experiment should characterize three related levels when he or she works with a child, assuming that the researcher expected a given level, say, level n, because of previous work with that child. If a child is at level n (defined as the dominant level of thinking), then level $n + 1$ is always a potential level and level $n - 1$ is always a possible fallback level. More exaggerated extensions or regressions are possible, say, for level $n + 2$ or $n - 2$, but such extensions or regressions have not often occurred in our work. This outcome represents important empirical support for the theoretical construct of the developmental progressions of LTs (Sarama & Clements, 2009b). A child at level n who exhibits thinking at level $n + 1$ is giving a sign that she or he is operating in the zone of proximal development at that time and therefore is already potentially in transition and particularly amenable to supports to better establish and generalize thinking at level $n + 1$. Our recommendation relates to the need for the researcher who provides instructional tasks and questions to concern herself or himself with the kind of thinking associated with three adjacent levels from the trajectory rather than just one level of thinking. Of course, we acknowledge that maintaining a perspective broad enough to encompass three levels of the trajectory at once will rarely be practical, but attempting such a perspective is worthwhile. This cyclic consideration of a child's thinking while going through the experiences of the teaching experiment promises a more integrated analytic set of data for each child's case report during the retrospective analysis process. We must always take these apparent levels as hypothetical and remember that the finding of leveled performance depends directly on the use of a specific task and the interaction among that task, the researcher, and the child (including characteristics beyond measurement competencies, such as a child who is a perfectionist; see Kerkman & Siegler, 1993). By engaging in design work that is based on using an LT, our findings are situated within our stated theoretical framework, hierarchical interactionalism; further, our own models about measurement concepts have defined the scope of our investigation and questioning patterns.

DESIGN CHALLENGES

The study reported here was a mixed-method multisite and multistate research project. We triangulated qualitative data from teaching episodes with quantitative data from cross-sectional assessment instruments. We next revisit our decisions and the outcomes of those decisions related to (a) task design, (b) teaching-experiment methodology, (c) cohort children's perceptions of our work with them, and (d) the triangulation of multiple data sources and sites.

Design Challenge 1: Designing Tasks

One design challenge for our team has been the problem of framing tasks to give contexts for measurement activity without oversimplifying measurement concepts. At the same time, we had to avoid obscuring the intended concept. Part of this challenge is addressing the need for reciprocal noticing, because children have frameworks for attending to tasks that differ markedly from those of the teacher or researcher (cf. Turner, Dominquez, Maldonado, & Empson, 2013). For our work on the research reported here, tasks—the third component of an LT—were necessary to elicit thinking and maybe encourage learning; however, the main purpose of this project was to evaluate and refine the developmental progressions, the second component of the LTs. Furthermore, these developmental progressions furnish excellent guides to those seeking to build more elaborated instructional tasks and descriptions of children's thinking and behaviors. Some tasks appeared to require more sophisticated thinking of the children than we expected. Others seemed to allow children to bypass the central question without the sophistication of measurement concepts that we had planned to engage.

The tasks also influenced the kind of interaction that occurred, because other tasks might have garnered other responses and hence different aspects of children's thinking and behaviors. Our selection of tasks characterizes our sense of the central conceptual aspects of spatial measurement, but we expect that other researchers may well contribute different tasks, questioning, and analysis that will extend or modify the literature on spatial measurement. Here we summarize five aspects of measurement education that we came to appreciate increasingly across the timeframe of our project:

1. Comparison is a fundamental conceptual basis for measurement: We wanted to encourage measurement rather than merely ask a child to "measure this," as textbooks often do. We began setting up comparison questions in which direct comparison was not possible for the child, by asking questions about a phenomenon that the child might find curious or relevant (e.g., Which containers should hold the most water or the least water? Can you arrange them in a line to show the smallest, next smallest, and the largest at the end?). Comparison questions are common elements of careful investigation, such as in science. We asked children to perform indirect perceptual comparisons at first and then comparisons that did not

have visual connections and that required a quantitative comparison with unit, as a partition of the whole. If the direct comparison was possible, then children did not need to measure, so we often worked to make direct comparison impossible or impractical.

2. Multiple representations are necessary to build a comprehensive understanding of measurement: During the third and fourth years of our project, we learned the importance of asking the children to develop connections among multiple representations as a means of engaging them in conceptual accounts of dimension, structuring space, and logical operations on continuous quantities. To support this development, we started asking the children to operate by coordinating their mental reasoning and a drawing within the same task. To prompt verbalizations of the children's mental actions on objects, we first asked the children to guess without using their hands. We then asked them to show how the units would cover the region by drawing or by using a drawing of a figure. Ideally, we asked them whether the figural representation fit well with their previous guess and whether they had imagined a structured region or space that matched what they produced when they worked with the figural image. This process helped children develop descriptive language and active words that they linked as anticipated outcomes (more advanced schemes), instead of merely asking them to measure something and then state the outcome.
3. Children need exposure to measurement in multiple contexts: We wanted to help children's thinking supersede any contexts in which researchers set tasks. We wanted to help them notice the essential role of units and structured collections of units. We therefore varied the context of tasks and our language to set the tasks. We checked whether children would identify the invariance of unit as it occurred across differing contexts. We wanted children to learn that units, once assigned, must relate to any other set of units or any further unit that one might assign within the same task or situation. For example, in the spring of Grade 3, the third semester, we presented children at the Midwest site with two rectangular prisms built with cubes of different sizes. The first prism was $2 \times 2 \times 4$ and consisted of 1-inch cubes; the second prism was $2 \times 3 \times 4$ and consisted of 1-centimeter cubes. The inch-cube prism was in fact much larger and appeared to be larger. To learn whether the children would follow perceptual impressions about volume or would count cubes in both prisms to compare volume (rather than appealing to a consistent unit across both objects), we told the children to consider the thinking of someone else, as in the classic Piagetian clinical interview. We asked them to suppose that another child had said that the figure consisting of the 1-centimeter cubes had a larger volume because it consisted of more cubes. We asked the children whether they agreed with that claim. If they agreed with the claim (an incorrect response), we asked the children to imagine that both structures consisted of ice and that the

ice melted, thereby changing both the context and the structural setting. Then we asked which would melt into more water. We found that the switch in context and even the substance to measure (from wooden blocks to ice cubes) often prompted children to conserve quantity and set the number of units along with the size of units into the background. Of course, our next challenge was to help the children reconnect their scheme for conserving ice or wood to their scheme for quantifying. Thus, we used context changes to help children abstract ideas about quantity, both through intuition and through reasoning with units.

4. Managing information in diagrams can prompt structural analysis: We made some design changes because of our goal of prompting thinking about quantity and unit identification for length and area tasks to help us identify children's structural insights and awareness. During the first year, when we presented an area-measurement task to children, we began designating side lengths by using varied symbols to show the linear units, prompting further verbal interaction with the children while they worked to measure. For example, we alternated the presentation of diagrams to include tick marks, dots, or number labels on one pair of orthogonal sides; in other cases, we presented number labels on all sides, we provided numbered rulers, or we furnished unnumbered rulers that had only tick marks. We also made purposeful decisions regarding the identification, location (inside or outside the region), and shape of the unit for area tasks. We varied these elements of the area-measure tasks to support children's engagement in structuring the measurement operation with area units and collections of units (Battista, Clements, Arnoff, Battista, & Borrow, 1998).
5. Providing children with large units may help them address struggles. During the first year, we wanted to help children value the process of measurement, including the value of precision; so on the advice of a participating classroom teacher, we started using large units to shift the scale of the workspace. During Grade 2, Sara, a child in the Midwest cohort who struggled on measurement tasks, had not attended to inconsistencies among reported lengths (perhaps she had not noticed them); and her classroom teacher suggested that we use large units. We found that Sara was less dismissive of errors made with a large unit than with a small unit, and she soon progressed to a more advanced level. For a similar reason, we also shifted the scale of the workspace during Year 4 work on volume with other children.

Design Challenge 2: Using a Teaching-Experiment Methodology

A second design challenge relates to the teaching-experiment methodology. Researchers may conduct teaching experiments within several different instructional situations: having a researcher work with an individual child, having a researcher work with another researcher acting as an observer or "witness" (Steffe & Thompson, 2000), having a researcher and a research witness working similarly with two children who collaborate, working with a group, or working with an entire classroom. The following is a brief historical review of our use of teaching experiments, predominantly using the first approach with individual children, in building and checking LTs. Teaching experiments require intensive resources for conducting and analyzing them immediately as well as retrospectively.

During the first year of our teaching-experiment work, we focused on identifying the characteristics of the early levels of the three developmental progressions by following individual children in the Midwest cohort in Grade 2. The records of these teaching episodes emphasize the attention to prior tasks and ways of extending those tasks to obtain increasingly sophisticated levels of thinking from each child. Each child's tasks differed from those of the other children and depended on their prior thinking and behaviors and the predictions of the research team. The records of our work with the Northeast cohort in pre-K emphasized the development of more consistent tasks across all the cohort children, showing relative responses across children. Each approach had drawbacks and benefits. Focusing our analysis on individual children in the Midwest cohort produced a more idiosyncratic narrative of developing knowledge and tended toward assessment rather than instructional support. In contrast, emphasizing task development spanning the group of children in the Northeast cohort produced a broader set of tasks that tended toward generalizable implications regarding instructional support and less toward differentiation of knowledge.

During the following year, both teams focused on presenting consistent tasks to all the children at the same grade level, thereby furnishing a better foundation for cross-site analysis of the LTs and the data from the teaching episodes. Our approach during Year 2 was closer to the activity of a classroom teacher and addressed learning opportunities through specific task sequences for an entire group (class) of children (e.g., the fixed-perimeter task sequence described in Chapter 2). This approach emphasized the assessment of children's thinking in consistent terms (setting a common metric), but the approach was less sensitive to variations in children's thinking and behaviors. To find out more about the individual cases, the research teams at both sites shifted the focus of task design once more. They began constructing tasks with varying levels of demand, as designed or predicted, to meet various levels of expressed sophistication for thinking in the spatial measurement domain, especially in the spring semester of the second year of the study. Each design decision led to shifts in our perspective. Nonetheless, we endeavored to maintain a consistent focus on building longitudinal accounts of the children's growing knowledge of measurement.

Design Challenge 3: Cohort Children's Perceptions of the Teaching Episodes

Finally, we consider some problematic aspects of our research methods, especially the teaching experiment as it relies on repeated cycles of teaching episodes with a researcher. To what extent do children respond in an attempt to give the researcher what he or she wants instead of following their natural progression of thinking? What did children in our study see as the expectations of the research team members? Children may have schematized the teaching episode context—thinking, "I won't get help, but I am supposed to keep working anyway" or "I have to work with blocks and things and answer a lot of questions" (cf. Ginsburg, 2009; Leinhardt & Putnam, 1987). At times, children would seem to think, "I want to draw to understand it"; however, in many other instances, the children appeared to think much differently: "I'm supposed to draw in these interviews."

We believe that notes about the interactive sets of expectations between researcher and subject are important for others who conduct extensive sets of teaching episodes with children in elementary school. The researcher and a child mutually constitute their interactions. Although a researcher who functions simultaneously as a teacher may be prompting the children to show their current way of interpreting an idea, the children may be attempting to have that teacher give them more clues than he or she gave at the outset of the task before they interpret the idea. In such situations, children may well respond with silence as a way to prompt the teacher to furnish further structure than the teacher had intended to provide. Brousseau (1997) described many related phenomena in teaching situations, including analyses of the negotiation of tasks, reduction of difficulty level, and funneling questions to prompt responses to difficult tasks. We found that reminding ourselves of these ideas was important for the ongoing work of predicting, checking, and supporting children's thinking from session to session.

For example, when we asked Anselm about his approach to a similar task during a previous teaching episode at the end of the spring semester during his Grade 2 year, he reverted to his previous behavior. It may have been his attempt to follow directions; from our standpoint, researchers should avoid giving children the impression of approval by an expert that occurs if one leaves a recently developed behavior to remain without any challenge. We wanted to learn whether Anselm would hold onto the new thinking pattern of counting gaps between hash marks as units of length or whether he would abandon the pattern and revert to a less sophisticated behavior of counting all hash marks along an object. Learning to respond to children without implying either agreement or disagreement with their behavior or statement proved demanding but essential to our work within a teaching experiment. Taking reflective notes and addressing common challenges may prove helpful. The researcher may use narrowing questions, for example, funneling questions, or differential questioning that is not based on prior findings with various subjects.

Design Challenge 4: Coordinating the Analysis of Data Across Sites

One of our goals was to coordinate our research analysis across different sites to prompt critical discussions of the analytic processing of the LT as a tool for describing children's thinking in the spatial measurement domain. By pressing research teams to explain and to furnish a rationale for coding decisions on a regular basis, greater opportunity existed to establish rigorous and justifiable claims that would represent the interpretation and theoretical commitments of researchers at both sites.

Qualitative data. Our research team worked to maintain consistency in our use of the LTs and in our accounts of the children in the two cohorts from year to year. However, annual changes in staffing, including multiple changes in the team of graduate research assistants at both research sites, complicated our situation. Our strategy for handling these demands shifted slightly from year to year, yet we did so primarily by keeping a constant focus on the theoretical accounts of children's developing schemes for the measurement content that the three LTs offered. Focusing on the schemes from the LTs gave us a theoretically coherent set of terms and descriptors for writing narratives about the progress of the children. It also provided natural checkpoints to verify whether team members were interpreting events through the LT in similar ways. The following sections give examples of the type of discussions that took place across sites to establish coding procedures for elements of our LT claims.

We found that the production of graphical displays of the developmental progression itself served as a practical way to coordinate our teams across sites and over the multiple years of the project. During the first year, we collaborated on displays of the accumulated findings from our work on the developmental progression of the LT for length measurement. The Northeast research team had displayed multiple levels for a given session to show varying kinds of thinking that seemed apparent during that teaching episode, but the Midwest research team had identified a predominant level for a given session and did not display the less dominant levels. The ensuing discussion about which approach to adopt led our combined group of researchers to examine both ideas: the need to identify a predominant level for that session and that child, as well as the complementary value of identifying peripheral, but less central, levels. We decided to present two separate displays of the accumulated findings, one per site, but from that point on to integrate our reporting to attend to a main level of thinking in addition to reporting less prominent levels that the student exhibited. We expected each researcher to identify the most prominent type of thinking and categorize it along the levels of the developmental progression of the LTs for each child and teaching episode.

Eventually, team members collaborated to establish a procedure for matching children's performance on tasks with the LT levels; the pattern emphasized framing task performance with a dominant level. After discussions across the two research sites, we also began including evidence of higher potential thinking or evidence of a fallback to a possible lower level during the second year of our study. These discussions arose when we attempted to report the performance of children over longer time periods in a brief poster presentation format after the first year of the project. Our analytical practice shifted: We began making level claims by each teaching episode (on a given day). We subsequently used these teaching-episode-level claims to make semester-level claims. We did not fully elaborate and completely implement this system of analysis and analyzing the teaching-experiment data according to the LT levels until the third year of the study. Therefore, we had to reanalyze the work done during the first 2 years of the teaching experiment after setting the comprehensive analytical approach.

During the retrospective analysis phase of the project, the coding of Drew's invented border-ring approach (see Figure 5.2) generated discussion both within and across sites. All members of the research team agreed that although Drew may have been thinking about groups of units in a way that would allow him to cover the region in a systematic way, this approach was unproductive because it would likely produce an incorrect numeric response. Discussion about the coding of instances in which Drew used his invented border-rings rule focused on whether his thinking was most consistent with the Primitive Coverer, Area Unit Relater and Repeater, or Partial Row Structurer level. The Northeast research team at first argued that the researchers should categorize this approach as being thinking at the Partial Row Structurer level because it showed that Drew was capable of using additive thinking with groups of units. Some members of the Midwest team that had been working with Drew and following the children in Grades 2 through 5 agreed that this invented border-ring rule constituted partial structuring (the Partial Row Structurer level) because it included using groups of units to cover space. However, other members of the team at this same site suggested that Drew's counting around the border and then attending to the interior units was more consistent with the description of the Primitive Coverer level of the LT for area measurement (see Figure 3.1). We resolved the dispute by determining that Drew was attempting to cover the space, although he did not identify a unit. He also did not explain how those numbers corresponded to locations in the region. Because of his purposeful attendance to covering the entire square and his struggle to explain exactly how and where 24, 12, 6, and 3 "centimeters" would fit, we determined that his thinking and behaviors were consistent with the Primitive Coverer level. This extended discussion prompted us to identify and reconcile differences in our interpretations of various components of the developmental progressions of the LTs.

Quantitative data. Although the focus of our study was qualitative, we also used quantitative data to provide additional support for our findings. In previous chapters, we highlighted the use of Rasch modeling to verify and extend the accounts of children's achievement in length, area, and volume measurement across a sample of more than 250 children. (For Rasch modeling, it is possible to establish stable parameter estimates with as few as 100 respondents; see Linacre,

1994; Reeve & Fayers, 2005.) For each LT, we established the unidimensionality of the construct (i.e., length, area, or volume) and presented evidence to demonstrate fit to the Rasch model. Each of the individual Rasch models therefore supported the developmental progression of that particular LT.

A lingering question that we still seek to answer is whether our assessment is of general or specific measurement competencies: Do the items from our assessment together target spatial measurement generally; or as we have presented in previous chapters, do they specifically target length, area, or volume? When we examined the entire pool of assessment items including length, area, and volume topics, the fit to the Rasch model was not as strong as it had been for the specific constructs taken separately (51.3% of raw variance explained by the measures with the unexplained variance in the first contrast at 2.3% with an eigenvalue of 2.4). Although still within acceptable ranges (Bond & Fox, 2007), length, area, and volume topics together might not form a single unidimensional trait.

Our choice to use Rasch modeling for our analyses lay in the strength that it offers when data fit the model, even though it only predicts one parameter. Because our combined data do not provide a strong fit to the Rasch model, we anticipate extending our quantitative analysis, possibly turning our attention to other statistical models. Multidimensional item response theory models might offer a better way to look at the length, area, and volume trajectories collectively, allowing for variation among the trajectories but highlighting commonalities. One drawback, though, is that these models require a much larger sample size. We are using the Q-matrix model and rule space methodology (Tatsuoka, 2009) in other projects (e.g., Clements, Sarama, & Tatsuoka, 2011), and this model might furnish another option, but it would necessitate recoding each of our items according to the specific attributes that they measured. We would possibly have to increase our item pool to provide a complete Q-matrix; and in addition, we would need a larger sample size.

Outside the statistical model, we also faced challenges in writing items of higher cognitive demand. In using written items on the assessments, as we did with the older children, viewing physical manipulations by the child, such as iteration of units, was impossible. For area, we could view children's drawings; but for length and volume, our items were lacking in this regard. Area—or volume—unit-iteration items may be more difficult for children than actually measuring an object. For volume, using a written assessment necessitated the use of two-dimensional representations for three-dimensional objects. This outcome opened the door to a similar problem in which volume-unit-iteration items would not reflect the actual measuring.

This result suggests a critical question that we want to answer in future work. Is our effort to understand children's thinking at the internal level—mental actions on objects—leading us to pose tasks and questions that are more demanding because they require immediate cognizance about actions and goals for those actions rather than implicit use of ideas or partially formed concepts? If so, we may need to continue our more theoretically motivated research without expecting that the procedures, tasks, and questioning routines can translate directly into standard classroom practices for teachers. However, we are also interested in bridging research and practice enough to keep the research agenda and activity relevant to standard classroom practice. For us, this outcome is a rationale to pursue questions of children's thinking by using research practices and processes that may rarely transfer directly into classroom activities and yet to pursue ways of translating our approaches into practical explorations of questioning routines, task descriptions, and assessment instruments that teachers may implement directly in everyday classroom work (e.g., Sztajn et al., 2012; Wilson, 2009).

Summary

Our main research question is, How do children's thinking and learning about each of the three domains of spatial measurement develop over time from pre-K through Grade 5? We found that the initial developmental progressions of research-based LTs were generally accurate representations of children's thinking and learning and thus were useful tools for predicting and analyzing children's behaviors. For each domain, the empirical evidence indicated productive revisions and enhancements to the LTs and resulted in three revised developmental progressions—described in detail in Chapters 2, 6, and 10—for length, area, and volume, respectively. The following are our reflections on this series of studies in the form of tentative recommendations for researchers who wish to further research LTs in geometric measurement or other mathematical domains.

- Studying LTs is a fecund approach to learning how children think and learn mathematics and how instructional tasks and teaching strategies can support their thinking and learning.
- Simultaneously considering an LT as a research tool and as a construct that researchers should refine requires careful reflection, which is aided by following a protocol. For example, the researcher should immediately document data inconsistent with the extant LT as disconfirming evidence and should check other cases to ascertain whether this case is typical or atypical. If typical or at least not idiosyncratic, modifying the initial LT level may be advisable.
- Changes to the LT can simply fill gaps or can make elaborations; these changes are useful. Other changes may imply a restructuring of the LT, with consideration of theoretical, empirical, and practical ramifications. Researchers should thoroughly check these changes with new data collection if possible; and after verifying the changes, researchers should document the changes and widely disseminate them.
- Combining cross-sectional and longitudinal methods furnishes complementary valuable perspectives and triangulation of findings.

IMPLICATIONS AND DIRECTIONS FOR FUTURE RESEARCH

We still have questions regarding the development of children's knowledge of geometric measurement, and we pose these questions and elaborate on the issues here. First, are the LT levels indivisible units of description for development of such knowledge, or should we study development by following components of the LTs? More analysis is necessary to relate cognitive components of the developmental progressions to the LTs themselves. We want to study components of developmental progressions, such as unit composition schemes, comparison schemes, or coordination schemes. Some researchers have argued that these ideas generalize directly throughout related domains of geometric measurement, namely, length, area, and volume (Curry, Mitchelmore, & Outhred, 2006). Our work suggests that children learn these ideas differently at times. We believe that researchers should further investigate longitudinal patterns for children so that researchers can explore the potential of a theory of measurement that might refine the learning of related domains. We expect that the interrelatedness of knowledge for length, area, and volume depends as well on progress along LTs for partitioning both number and space, composing and decomposing space, and arithmetic operations on number. These far-reaching questions deserve further study.

One might also separate the study of knowledge from the study of reasoning by looking at stable patterns of sophistication in children's reasoning with tools, language, and principles across domains such as spatial measurement or rational numbers, especially fractions. In contrast, one might attempt to focus on the content knowledge and leave logical reasoning in the background, although researchers rarely maintain this distinction. We note that several recommendations for curriculum and research programs emphasize the importance of integrating thinking, behavior development, conceptual understanding, disposition toward mathematics, and procedural fluency (Kilpatrick, Swafford, & Findell, 2001; National Governors Association Center for Best Practices [NGA] & Council of Chief State School Officers [CCSSO], 2010). We tried to consider both content knowledge and logical reasoning in this study, but there may be advantages to isolating them.

Second, is it important to work from natural contexts for examining children's development of spatial measurement schemes? And is the teaching-experiment methodology relevant to the way that children learn and develop within a traditional classroom setting? Our experiences in classroom situations have been consistent with our individual teaching-experiment work (cf. Stephan, Bowers, Cobb, & Gravemeijer, 2004), but we believe the advantage of classroom study (e.g., a classroom teaching experiment) would be a more natural opportunity to directly relate our instructional experiences to classroom teachers by sharing journal notes and video records of our work in larger classroom contexts.

Third, will these LTs generalize well? Our application of the LTs has centered on specific teaching episodes, and our cohorts of children from two sites in the United States have limited the application of the LTs. We were focusing on the links from one level of the LT to the next level and giving accounts of the mechanisms for moving from one level in an LT to the next while we developed sequences of lessons. We maintained a design focus on the children's sequence of learning to think with units and then with groups of units, with an emphasis on comparative reasoning. It would be helpful to modify these conditions and check for the generalizability of the LTs across culture, socioeconomic identity, and subsets of learner characteristics, including struggling learners or advanced learners at any given grade level in school.

Fourth, do the tasks that we used in this study determine children's performance to some degree, or did they serve as broadly representative interventions? We designed the task sequences at each successive level as a way to address the specific mental actions on objects that are characteristic of a given level. We expected that the tasks clarified and extended the theoretical description of that level in the trajectory. We did not expect the tasks that we used to influence the level of performance for any child, but rather to make the level of performance evident. In the end, after collecting and analyzing 4 years of data, we found that tasks can be either problematic or supportive, can emphasize a critical insight, or can overwhelm a child and cause the child to fall back or "fold back" (cf. Pirie & Kieren, 1994) to an established level of thinking and reconsider a difficult task from this more confident position. Most of the time, the tasks that we used yielded an assessment of children's thinking. In some situations, we presented tasks and gave scaffolds designed to help children build further knowledge, such as those presented during Measurement Club at the Northeast site in Grade 2 or in the classroom teaching episodes used in Grade 1 at the Northeast site and in Grades 3, 4, and 5 at the Midwest site. These types of tasks proved challenging to engineer but were more useful for both assessment and instruction.

Fifth, we still want to learn whether we can design sequences of tasks to prompt growth beyond a single adjacent level. That is, can we set instructional sequences of tasks and reflective discussions that carry critical structures of spatial measurement further than the next level of thinking, thereby allowing children to skip levels? If so, what does that ability to skip levels imply about our theoretical claim for levels of thinking that are sequential? Further work is necessary, and it should involve a wide range of task sequences that support such transitions, allowing a generalizable answer. We briefly describe our future work in the final paragraph of this chapter.

IMPLICATIONS AND DIRECTIONS FOR EDUCATIONAL PRACTICE

We believe that our findings have useful implications for education praxis as well, including writing and revising standards, policy documents, curriculum design, and teaching. Criticism of the *Common Core State Standards for Mathematics* (NGA & CCSSO, 2010) and present curricula include a concern that they fail to emphasize geometric measurement (e.g., J. P. Smith, Males, Dietiker, Lee, &

Mosier, 2013). The research that we review and present here may support productive revisions. In general, the revised developmental progressions for measurement LTs (see Chapters 2, 6, and 10) give detailed empirically based guidance for developing standards and curricula as well as for teaching, especially using the powerful technique of formative assessment. To use formative assessments, teachers must ask themselves the following questions (Shepard, Hammerness, Darling-Hammond, & Rust, 2005):

- What is the key error? (Although we use the term *error* directly here, we note that errors may be partial conceptions or misconceptions. As such, they often constitute productive attempts that we should address directly rather than dismiss.)
- What is the probable reason that the child made this error?
- How can I guide the child to avoid this error in the future?

LTs furnish a tool that helps teachers answer these questions. Understanding the goal of the LT may help the teacher identify an error in mathematical thinking. Understanding the developmental progression of the LT allows the teacher to understand the level of thinking and the reason that the student made the error. Understanding tasks relevant to each of the levels of the LT furnishes information that the teacher can use to guide the child's thinking and learning. Teachers need support to revise tasks along an LT to challenge children during classroom instruction processes (Wickstrom, 2014). The LT can support professional knowledge of children's thinking by furnishing related versions of tasks suited to the range of levels of thinking in any one classroom. For example, Empson (2011) emphasized the importance of bottom-up instructional exchanges between teachers and children, whereby the teacher begins from the children's specific thinking and finds ways to extend it that follow closely along the child's initial direction and emphasis. This emphasis is a specific extension of the findings and recommendations of the Cognitively Guided Instruction research program (Carpenter, Fennema, Peterson, Chiang, & Loef, 1990; Fennema, Carpenter, & Franke, 1992).

We concur that instructional tasks that researchers can associate with each level of the LT can be important tools for teachers, but much work remains to find ways of helping teachers incorporate collections of tasks into their teaching practices. Our observations about the approaches of specific children should encourage further attention to the variation and breadth of strategies among children who are learning about measurement. We believe that LTs give teachers a way of looking for variability among learners and finding ways to reflect the wide array of learning paths to their own children while they work with their class. Below we report our reflections on curriculum development and teaching.

- Just because children can read a number from a ruler does not mean that they understand that the number represents an iteration of equal-length units, as children do at or above the End-to-End Length Measurer level.

- The initial LT for length was generally valid. One alteration was that direct comparison is an important foundation for both indirect length comparison and measurement (e.g., end to end) but that indirect length comparison is not a prerequisite to measurement. It instead appears to develop somewhat independently of and in parallel with the levels of measurement.
- Children who could measure end to end benefited from using larger units (a stride or step-size) and seeing demonstrations of ineffective or incorrect strategies in exaggerated ways by using these units, as well as from making drawings of successive positions of an iterated unit and using sweeping motions to record unit lengths between hash marks.
- Children who could measure by iterating a physical unit benefited from instruction that combined counting of intervals, counting hash marks between intervals, and attending to sequential displays of number labels at the hash marks.
- Children at a young age can tile a region but do not necessary know how to draw, conceptualize, or quantify that array.
- Just because children can report the number of square units necessary to cover a rectilinear region does not mean that they understand that the number of square units is a measure of area (i.e., a quantification of "taking up space"). Asking children to measure with nonsquare units can support refinements in conceptual knowledge of the space as a quantity.
- Just because children can count the number of unit cubes in a prism does not mean that they understand that the number of unit cubes is a measure of volume (i.e., a quantification of "taking up space").
- Children often can succeed early at tasks that deal with capacity by treating it psychologically (as a one-dimensional entity filling a graduated cylinder, reading the measure along its length). Learning to understand its three-dimensional character mathematically and relating filling and packing volume are subsequent important achievements that mathematics curricula and teaching often ignore.
- Children often operate on measures of area and measures of volume without necessarily distinguishing the two measures as operations with different dimensions.
- Finally, can children employ formulaic descriptions of general measuring solutions before they grasp the related operations on units and groups of units? For example, children struggle to integrate a formula for the volume of a prism or a cylinder (area of the base × altitude of the prism) with a formula for the volume of a rectangular prism (length × width × height).

We believe that these suggestions may be useful. But we also believe that no simple list of recommendations can adequately represent what is potentially valuable in this research for those working on standards, curriculum, professional development, and teaching. Again, the developmental progressions of the LTs for length, area, and volume measurement include these suggestions and other insights and implications, situating them within an empirically supported organization.

FUTURE DIRECTIONS FOR RESEARCH

Our findings suggest several unanswered questions for future research. The first unanswered question involves whether vocabulary developments may be important, even necessary, for children to construct higher levels of length understanding. Vocabulary development seems especially important at the first level, Length Quantity Recognizer, for which the initial LT specified the use of appropriate vocabulary to identify length or distance. Most children, whose development and appropriation of length vocabulary scaffolded their learning (cf. Vygotsky, 1934/1986), realized this expectation. The basis for at least some children's responses to some tasks, however, appeared to be theorems-in-action (Vergnaud, 1982) without observed verbal mediation. This absence of vocabulary in processing measurement tasks did not appear to hinder children's development of new solution strategies or at least some initial new levels of thinking. We do not consider this outcome to be a finding, because the small number of participants and lack of methodology focusing on language limit possible conclusions. Instead, we suggest that researchers might address this issue in future research, such as studies of the relationship between language competencies and cognition (e.g., Praet, Titeca, Ceulemans, & Desoete, 2013; Purpura & Ganley, 2014; Schumacher & Fuchs, 2012; Vukovic & Lesaux, 2013) or studies that examine the interactions around teaching and learning by using a sociolinguistic approach (e.g., Kaartinen & Kumpulainen, 2012; Moschkovich, 2007, 2010).

A second unanswered question involves the implicit expectation that children conceptualize measures of length as comparative operations or ratios. Researchers have expected to find this conceptualization while they investigate children's length-measurement activity, because of the fundamental importance of ratio and comparative reasoning in both science and mathematics, but more work is necessary to clarify children's concepts for comparison in relation to their concepts for unit and unit iteration.

Finally, more exploratory work across all three measurement domains is advisable with children in the middle school grades as a way of extending the LT levels to account for more sophisticated levels of thinking. Continued exploration of the instructional tasks for pre-K through Grade 5 is also necessary to give a more generalized account of the mechanisms by which children shift from one level to the next level of thinking about length, along with further research to extend our work to more diverse populations.

In this monograph, we described, in detail, revisions to the developmental progressions of the LTs for length, area, and volume measurement, with only occasional notes on instructional tasks. We are addressing this limitation for curriculum and teaching in our most recent National Science Foundation–funded project, titled *Learning Trajectories to Support the Growth of Measurement Knowledge: Pre-K Through Middle School* (Grant No. DRL-1222944). On this project, we not only continue to use the methods that we found productive and reported on in this monograph, but also evaluate the instructional tasks more completely and more rigorously by using microgenetic analyses (Siegler & Svetina, 2006). We look forward to interacting with colleagues regarding the findings presented herein as well as their suggestions for our work on the new project.

References

American Association for the Advancement of Science. (2001). *Atlas of science literacy*. Washington, DC: American Association for the Advancement of Science & National Science Teachers Association.

Baroody, A. J., Lai, M.-L., & Mix, K. S. (2006). The development of young children's number and operation sense and its implications for early childhood education. In B. Spodek & O. N. Saracho (Eds.), *Handbook of research on the education of young children* (pp. 187–221). Mahwah, NJ: Erlbaum.

Barrett, J. E., & Battista, M. T. (2014). Two approaches to describing the development of students' reasoning about length: A case study for coordinating related trajectories. In A. P. Maloney, J. Confrey, & K. Nguyen (Eds.), *Learning over time: Learning trajectories in mathematics education* (pp. 97–124). Charlotte, NC: Information Age Publishing.

Barrett, J. E., Clements, D. H., Klanderman, D., Pennisi, S.-J., & Polaki, M. V. (2006). Students' coordination of geometric reasoning and measuring strategies on a fixed perimeter task: Developing mathematical understanding of linear measurement. *Journal for Research in Mathematics Education, 37*(3), 187–221.

Barrett, J. E., Cullen, C., Sarama, J., Clements, D. H., Klanderman, D., Miller, A. L., & Rumsey, C. (2011). Children's unit concepts in measurement: A teaching experiment spanning grades 2 through 5. *ZDM: The International Journal on Mathematics Education, 43*(5), 637–650. doi:10.1007/s11858-011-0368-8

Barrett, J. E., Sarama, J., Clements, D. H., Cullen, C., McCool, J., Witkowski-Rumsey, C., & Klanderman, D. (2012). Evaluating and improving a learning trajectory for linear measurement in elementary grades 2 and 3: A longitudinal study. *Mathematical Thinking and Learning, 14*(1), 28–54. doi:10.1080/10986065.2012.625075

Bartsch, K., & Wellman, H. M. (1988). Young children's conception of distance. *Developmental Psychology, 24*(4), 532–541. doi:10.1037/0012-1649.24.4.532

Battista, M. T. (2012). *Cognition-based assessment and teaching of geometric measurement: Building on students' reasoning*. Portsmouth, NH: Heinemann.

Battista, M. T., & Clements, D. H. (1996). Students' understanding of three-dimensional rectangular arrays of cubes. *Journal for Research in Mathematics Education, 27*(3), 258–292. doi:10.2307/749365

Battista, M. T., & Clements, D. H. (1998). Students' understanding of three-dimensional cube arrays: Findings from a research and curriculum development project. In R. Lehrer & D. Chazan (Eds.), *Designing learning environments for developing understanding of geometry and space* (pp. 227–248). Mahwah, NJ: Erlbaum.

Battista, M. T., Clements, D. H., Arnoff, J., Battista, K., & Borrow, C. V. A. (1998). Children's spatial structuring of 2D arrays of squares. *Journal for Research in Mathematics Education, 29*(5), 503–532. doi:10.2307/749731

Bausano, M. K., & Jeffrey, W. E. (1975). Dimensional salience and judgments of bigness by three-year-old children. *Child Development, 46*(4), 988–991. doi:10.2307/1128410

Beilin, H. (with Klein, A., & Whitehurst, B.). (1982). *Strategies and structures in understanding geometry* (Research Report No. NSF/SED-82020). Retrieved from ERIC database. (ED225808)

Bond, T. G., & Fox, C. M. (2007). *Applying the Rasch model: Fundamental measurement in the human sciences* (2nd ed.). Mahwah, NJ: Erlbaum.

Bragg, P., & Outhred, L. (2004). A measure of rulers—The importance of units in a measure. In M. Hoines & A. Fuglestad (Eds.), *Proceedings of the 28th Conference of the International Group for the Psychology of Mathematics Education* (Vol. 2, pp. 159–166). Bergen, Norway. Retrieved from http://files.eric.ed.gov/fulltext/ED489632.pdf

Brousseau, G. (1997). *Theory of didactical situations in mathematics* (N. Balacheff, M. Cooper, R. Sutherland, & V. Warfield, Eds. & Trans.). Dordrecht, the Netherlands: Kluwer.

Buss, A. T., & Spencer, J. P. (2014). *The emergent executive: A dynamic field theory of the development of executive function* [Monograph]. *Society for Research in Child Development, 79*(2).

Cannon, P. L. (1992). Middle grade students' representations of linear units. In W. Geeslin & K. Graham (Eds.), *Proceedings of the 16th Conference of the International Group for the Psychology of Mathematics Education* (Vol. 1, pp. 105–112). Durham, NH. Retrieved from http://files.eric.ed.gov/fulltext/ED383538.pdf

Carpenter, T. P., Fennema, E., Peterson, P. L., Chiang, C.-P., & Loef, M. (1990). Using knowledge of children's mathematics thinking in classroom teaching: An experimental study. *American Educational Research Journal, 26*(4), 499–531. doi:10.3102/00028312026004499

Carpenter, T. P., Franke, M. L., Jacobs, V. R., Fennema, E., & Empson, S. B. (1998). A longitudinal study of invention and understanding in children's multidigit addition and subtraction. *Journal for Research in Mathematics Education, 29*(1), 3–20. doi:10.2307/749715

Case, R., & Okamoto, Y. (with Griffin, S., McKenough, A., Bleiker, C., Henderson, B., & Stephenson, K. M.). (1996). The role of central conceptual structures in the development of children's thought [Monograph]. *Society for Research in Child Development, 61*(1–2). doi:10.2307/1166077

Cavanagh, M. (2008). Reflections on measurement and geometry: Area measurement in year 7. *Reflections, 33*(1), 55–58. Retrieved from http://www.curriculumsupport.education.nsw.gov.au/secondary/mathematics/assets/pdf/s4_teach_ideas/area/area_meas.pdf

Chappell, M. F., & Thompson, D. R. (1999a). Modifying our questions to assess students' thinking [Reflections on practice]. *Mathematics Teaching in the Middle School, 4*(7), 470–474.

Chappell, M. F., & Thompson, D. R. (1999b). Perimeter or area? Which measure is it? [Take time for action]. *Mathematics Teaching in the Middle School, 5*(1), 20–23.

Clarke, D. M., Cheeseman, J., Gervasoni, A., Gronn, D., Horne, M., McDonough, A., . . . Rowley, G. (2002, February). *Early numeracy research project final report.* Melbourne, Australia: Victorian Department of Education, Employment and Training, the Catholic Education Office, and Association of Independent Schools.

Clements, D. H. (1999a). "Concrete" manipulatives, concrete ideas. *Contemporary Issues in Early Childhood, 1*(1), 45–60. doi:10.2304/ciec.2000.1.1.7

Clements, D. H. (1999b). Teaching length measurement: Research challenges. *School Science and Mathematics, 99*(1), 5–11. doi:10.1111/j.1949-8594.1999.tb17440.x

Clements, D. H. (Ed). (2003). *Learning and teaching measurement.* 2003 yearbook of the National Council of Teachers of Mathematics. Reston, VA: National Council of Teachers of Mathematics.

Clements, D. H. (2007). Curriculum research: Toward a framework for research-based curricula. *Journal for Research in Mathematics Education, 38*(1), 35–70.

Clements, D. H., Agodini, R., & Harris, B. (2013, September). *Instructional practices and student math achievement: Correlations from a study of math curricula* (NCEE Evaluation Brief). Washington, DC: Institute of Education Sciences: National Center for Education Evaluation and Regional Assistance.

Clements, D. H., & Battista, M. T. (with Sarama, J.). (2001). *Logo and geometry.* Journal for Research in Mathematics Education monograph series (Vol. 10). Reston, VA: National Council of Teachers of Mathematics. doi:10.2307/749924

Clements, D. H., & Sarama, J. (2004). Learning trajectories in mathematics education. *Mathematical Thinking and Learning, 6*(2), 81–89. doi:10.1207/s15327833mtl0602_1

Clements, D. H., & Sarama, J. (2007). Early childhood mathematics learning. In F. K. Lester Jr. (Ed.), *Second handbook of research on mathematics teaching and learning* (Vol. 1, pp. 461–555). Charlotte, NC: Information Age Publishing.

Clements, D. H., & Sarama, J. (2009). *Learning and teaching early math: The learning trajectories approach.* New York, NY: Routledge.

Clements, D. H., & Sarama, J. (2014a). *Learning and teaching early math: The learning trajectories approach* (2nd ed.). New York, NY: Routledge.

Clements, D. H., & Sarama, J. (2014b). Learning trajectories: Foundations for effective, research-based education. In A. P. Maloney, J. Confrey, & K. H. Nguyen (Eds.), *Learning over time: Learning trajectories in mathematics education* (pp. 1–30). Charlotte, NC: Information Age Publishing.

Clements, D. H., Sarama, J., & DiBiase, A.-M. (Eds.). (2004). *Engaging young children in mathematics: Standards for early childhood mathematics education.* Mahwah, NJ: Erlbaum.

Clements, D. H., & Stephan, M. (2004). Measurement in preK to grade 2 mathematics. In D. H. Clements, J. Sarama, & A.-M. DiBiase (Eds.), *Engaging young children in mathematics: Standards for early childhood mathematics education* (pp. 299–317). Mahwah, NJ: Erlbaum.

Cobb, P., Confrey, J., diSessa, A., Lehrer, R., & Schauble, L. (2003). Design experiments in educational research. *Educational Researcher, 32*(1), 9–13. doi:10.3102/0013189X032001009

Cobb, P., & Gravemeijer, K. (2008). Experimenting to support and understand learning processes. In A. E. Kelly, R. A. Lesh, & J. Y. Baek (Eds.), *Handbook of design research methods in education: Innovations in science, technology, engineering, and mathematics learning and teaching* (pp. 68–95). New York, NY: Routledge.

Confrey, J., & Kazak, S. (2006). A thirty-year reflection on constructivism in mathematics education in PME. In A. Gutiérrez & P. Boero (Eds.), *Handbook of research on the psychology of mathematics education: Past, present, and future* (pp. 305–345). Rotterdam, the Netherlands: Sense.

Confrey, J., & LaChance, A. (2000). Transformative teaching experiments through conjecture-driven research design. In A. E. Kelly & R. A. Lesh (Eds.), *Handbook of research design in mathematics and science education* (pp. 231–266). Mahwah, NJ: Erlbaum.

Confrey, J., Rupp, A. A., Maloney, A. P., & Nguyen, K. H. (2012). *Toward developing and representing a novel learning trajectory for equipartitioning: Literature review, diagnostic assessment design, and explanatory item response modeling.* Unpublished manuscript, College of Education, North Carolina State University, Raleigh, NC.

Cullen, C. J., & Barrett, J. E. (2010). Strategy use indicative of an understanding of units of length. In M. F. Pinto & T. F. Kawasaki (Eds.), *Proceedings of the 34th Conference of the International Group for the Psychology of Mathematics Education* (Vol. 2, pp. 281–288). Belo Horizonte, Brazil.

Cullen, C. J., Miller, A. L., Barrett, J. E., Clements, D. H., & Sarama, J. (2011). Unit eliciting task structures: Verbal prompts for comparative measures. In B. Ubez (Ed.), *Proceedings of the 35th Conference of the International Group for the Psychology of Mathematics Education* (Vol. 2, pp. 249–256). Ankara, Turkey.

Cullen, C. J., Miller, A. L., Witkowski-Rumsey, C., Barrett, J. E., & Sarama, J. (2011, April). *Area hypothetical learning trajectory: Relating square units to non-rectilinear regions.* Paper presented at the Annual Meeting of the American Educational Research Association, New Orleans, LA.

Cuneo, D. O. (1980). A general strategy for quantity judgments: The height + width rule. *Child Development, 51*(1), 299–301. doi:10.2307/1129629

Curry, M., Mitchelmore, M., & Outhred, L. (2006, July). Development of children's understanding of length, area, and volume measurement principles. In J. Novotná, H. Moraová, M. Krátká, & N. Stehlíková (Eds.), *Proceedings of the 30th Conference of the International Group for the Psychology of Mathematics Education* (Vol. 2, pp. 377–384). Prague, Czech Republic.

Curry, M., & Outhred, L. (2005). Conceptual understanding of spatial measurement. In P. Clarkson et al. (Eds.), *Building connections: Research, theory and practice—Proceedings of the 28th Annual Conference of the Mathematics Education Research Group of Australasia* (pp. 265–272). Melbourne, Australia. Retrieved from http://www.merga.net.au/documents/RP262005.pdf

Daro, P., Mosher, F. A., & Corcoran, T. (with Barrett, J., Battista, M., Clements, D., Confrey, J., Daro, V., Maloney, A., . . . Sarama, J.) (2011, January). *Learning trajectories in mathematics: A foundation for standards, curriculum, assessment, and instruction* (CPRE Research Report No. RR-68). Philadelphia, PA: Consortium for Policy Research in Education. doi:10.12698/cpre.2011.rr68

Davydov, V. V. (Ed.). (1991). Psychological abilities of primary school children in learning mathematics. In L. P. Steffe (Volume Ed.), *Soviet Studies in Mathematics Education: Vol. 6* (J. Teller, Trans.). Reston, VA: National Council of Teachers of Mathematics.

Ellis, S., Siegler, R. S., & Van Voorhis, F. E. (2003, April). *Developmental changes in children's understanding of measurement procedures and principles.* Paper presented at the Biennial Meeting of the Society for Research in Child Development, Tampa, FL.

Empson, S. B. (2011). On the idea of learning trajectories: Promises and pitfalls. *Montana Mathematics Enthusiast, 8*(3), 571–598. Retrieved from http://www.math.umt.edu/TMME/vol8no3/Empson_TME_2011_article5_pp.571_596.pdf

Fabricius, W. V., & Wellman, H. M. (1993). Two roads diverged: Young children's ability to judge distances. *Child Development, 64*(2), 399–414. doi:10.2307/1131258

Fennema, E., Carpenter, T. P., & Franke, M. L. (1992). Cognitively guided instruction. *National Center for Reasearch in Mathematical Sciences Education Research Review: The Teaching and Learning of Mathematics, 1*(2), 5–9. Retrieved from http://files.eric.ed.gov/fulltext/ED372929.pdf

Gagné, R. M. (1970). *The conditions of learning and theory of instruction.* New York, NY: Holt, Rinehart, & Winston. (Original work published 1965)

Ginsburg, H. P. (2009). The challenge of formative assessment in mathematics education: Children's minds, teachers' minds. *Human Development, 52*(2), 109–128. doi:10.1159/000202729

Greeno, J. G., Riley, M. S., & Gelman, R. (1984). Conceptual competence and children's counting. *Cognitive Psychology, 16*(1), 94–143. doi:10.1016/0010-0285(84)90005-7

Hiebert, J. (1981). Cognitive development and learning linear measurement. *Journal for Research in Mathematics Education, 12*(3), 197–211. doi:10.2307/748928

Hiebert, J. (1984). Why do some children have trouble learning measurement concepts? *Arithmetic Teacher, 31*(7), 19–24.

Hirstein, J., Lamb, C. E., & Osborne, A. (1978). Student misconceptions about area measure. *Arithmetic Teacher, 25*(6), 10–16.

Izsák, A., Tillema, E., & Tunç-Pekkan, Z. (2008). Teaching and learning fraction addition on number lines. *Journal for Research in Mathematics Education, 39*(1), 33–62.

Kaartinen, S., & Kumpulainen, K. (2012). The emergence of mathematizing as a culture of participation in the early childhood classroom. *European Early Childhood Education Research Journal, 20*(2), 263–281. doi:10.1080/1350293x.2012.681136

Kamii, C., & Clark, F. B. (1997). Measurement of length: The need for a better approach to teaching. *School Science and Mathematics, 97*(3), 116–121. doi:10.1111/j.1949-8594.1997.tb17354.x

Kamii, C., & Kysh, J. (2006). The difficulty of "length × width": Is a square the unit of measurement? *Journal of Mathematical Behavior, 25*(2), 105–115. doi:10.1016/j.jmathb.2006.02.001

Kara, M., Eames, C. L., Miller, A. L., Cullen, C. J., & Barrett, J. E. (2011). Developing an understanding of area measurement concepts with triangular units. In L. R. Wiest & T. Lamberg (Eds.), *Proceedings of the 33rd Annual Meeting of the North American Chapter of the International Group for the Psychology of Mathematics Education* (pp. 1015–1023). Reno, NV.

Karmiloff-Smith, A. (1992). *Beyond modularity: A developmental perspective on cognitive science.* Cambridge, MA: Massachusetts Institute of Technology.

Kerkman, D. D., & Siegler, R. S. Individual differences and adaptive flexibility in lower-income children's strategy choices. *Learning and Individual Differences, 5*(2), 113–136. doi:10.1016/1041-6080(93)90008-G

Kieren, T., & Pirie, S. (1992). The answer determines the question: Interventions and the growth of mathematical understanding. In W. Geeslin & K. Graham (Eds.), *Proceedings of the 16th Conference of the International Group for the Psychology in Mathematics Education* (Vol. 2, pp. 1–8). Durham, NH.

Kilpatrick, J., Swafford, J., & Findell, B. (Eds.). (2001). Adding it up: Helping children learn mathematics. Washington, DC: National Academy Press.

Kordaki, M., & Potari, D. (2002). The effect of area measurement tools on student strategies: The role of a computer microworld. *International Journal of Computers for Mathematical Learning, 7*(1), 65–100. doi:10.1023/A:1016051411284

Lehrer, R. (2003). Developing understanding of measurement. In J. Kilpatrick, W. G. Martin, & D. Schifter (Eds.), *Research companion to* principles and standards for school mathematics (pp. 179–192). Reston, VA: National Council of Teachers of Mathematics.

Lehrer, R., Jacobson, C., Thoyre, G., Kemeny, V., Strom, D., Horvath, J., . . . Koehler, M. (1998). Developing understanding of geometry and space in the primary grades. In R. Lehrer & D. Chazan (Eds.), *Designing learning environments for developing understanding of geometry and space* (pp. 169–200). Mahwah, NJ: Erlbaum.

Lehrer, R., Jaslow, L., & Curtis, C. L. (2003). Developing an understanding of measurement in the elementary grades. In D. H. Clements (Ed.), *Learning and teaching measurement,* 2003 yearbook of the National Council of Teachers of Mathematics (pp. 100–121). Reston, VA: National Council of Teachers of Mathematics.

Lehrer, R., Jenkins, M., & Osana, H. (1998). Longitudinal study of children's reasoning about space and geometry. In R. Lehrer & D. Chazan (Eds.), *Designing learning environments for developing understanding of geometry and space* (pp. 137–168). Mahwah, NJ: Erlbaum.

Lehrer, R., & Schauble, L. (2000). Inventing data structures for representational purposes: Elementary grade students' classification models. *Mathematical Thinking and Learning, 2*(1–2), 51–74. doi:10.1207/S15327833MTL0202_3

Lehrer, R., & Schauble, L. (Eds.). (2002). *Investigating real data in the classroom: Expanding children's understanding of math and science.* New York, NY: Teachers College Press.

Lehrer, R., & Schauble, L. (2004). Modeling natural variation through distribution. *American Educational Research Journal, 41*(3), 635–679. doi:10.3102/00028312041003635

Leinhardt, G., & Putnam, R. T. (1987). The skill of learning from classroom lessons. *American Educational Research Journal, 24*(4), 557–587. doi:10.3102/00028312024004557

Linacre, J. M. (1994). Sample size and item calibration stability. *Rasch Measurement Transactions, 7*(4), 328. Retrieved from http://www.rasch.org/rmt/rmt74m.htm

Linacre, J. M. (2005). *A user's guide to Winsteps/Ministep Rasch-model computer program* (Version 3.57). Chicago, IL: Winsteps.com.

Linacre, J. M. (2014). *A user's guide to Winsteps/Ministep Rasch-model computer program.* Chicago, IL: Winsteps.com.

Lindquist, M. M., & Kouba, V. L. (1989). Measurement. In M. M. Lindquist (Ed.), *Results from the fourth mathematics assessment of the national assessment of educational progress* (pp. 35–43). Reston, VA: National Council of Teachers of Mathematics.

Maloney, A. P., & Confrey, J. (2010, July). *The construction, refinement, and early validation of the equipartitioning learning trajectory.* Paper presented at the 9th International Conference of the Learning Sciences, Chicago, IL.

Maloney, A. P., Confrey, J., & Nguyen, K. H. (Eds.). (2014). *Learning over time: Learning trajectories in mathematics education.* Charlotte, NC: Information Age Publishing.

Maratsos, M. P. (1973). Decrease in the understanding of the word "big" in preschool children. *Child Development, 44*(4), 747–752. doi:10.2307/1127719

Mariotti, M. A. (2002). The influence of technological advances on students' mathematical learning. In L. D. English (Ed.), *Handbook of international research in mathematics education* (pp. 695–723). Mantua, NJ: National Council of Teachers of Mathematics.

Miller, A. L. (2013). *Investigating conceptual, procedural, and intuitive aspects of area measurement with non-square area units* (Doctoral dissertation). Available from ProQuest Dissertations and Theses database. (Order No. 3572989)

Miller, A. L., Kara, M., Eames, C. L., Cullen, C. J., & Barrett, J. E. (2012). A comparison of three students' responses to area invariance tasks across grades 2–5. In L. R. Van Zoest, J. Lo, & J. L. Kratky (Eds.), *Proceedings of the 34th Annual Meeting of the North American Chapter of the International Group for the Psychology of Mathematics Education* (pp. 1042). Kalamazoo, MI.

Miller, K. F. (1989). Measurement as a tool for thought: The role of measuring procedures in children's understanding of quantitative invariance. *Developmental Psychology, 25*(4), 589–600. doi:10.1037/0012-1649.25.4.589

Miller, K. F., & Baillargeon, R. (1990). Length and distance: Do preschoolers think that occlusion brings things together? *Developmental Psychology, 26*(1), 103–114. doi:10.1037/0012-1649.26.1.103

Minsky, M. (1986). *The society of mind.* New York, NY: Simon & Schuster.

Moschkovich, J. (2007). Examining mathematical discourse practices. *For the Learning of Mathematics, 27*(1), 24–30.

Moschkovich, J. N. (2010). Language(s) and mathematics: Resources, challenges, and issues for reasearch. In J. N. Moschkovich (Ed.), *Language and mathematics education: Multiple perspectives and directions for research* (pp. 1–28). Charlotte, NC: Information Age Publishing.

Mullet, E., & Paques, P. (1991). The height + width = area of a rectangle rule in five-year-olds: Effects of stimulus distribution and graduation of the response scale. *Journal of Experimental Child Psychology, 52*(3), 336–343. doi:10.1016/0022-0965(91)90067-3

Mulligan, J., Prescott, A., Mitchelmore, M. C., & Outhred, L. (2005). Taking a closer look at young students' images of area measurement. *Australian Primary Mathematics Classroom, 10*(2), 4–8.

Murata, A., & Fuson, K. C. (2006). Teaching as assisting individual constructive paths within an interdependent class learning zone: Japanese first graders learning to add using 10. *Journal for Research in Mathematics Education, 37*(5), 421–456.

National Assessment of Educational Progress. (1983, April). *The third national mathematics assessment: Results, trends, and issues* (Report No. 13-MA-01). Denver, CO: Education Commission of the States. Retrieved from ERIC database. (ED228049)

National Assessment of Educational Progress. (2013). *NAEP data explorer.* Retrieved from http://nces.ed.gov/nationsreportcard/naepdata/report.aspx

National Council of Teachers of Mathematics. (2006). *Curriculum focal points for prekindergarten through grade 8 mathematics: A quest for coherence.* Reston, VA: Author.

National Governors Association Center for Best Practices & Council of Chief State School Officers. (2010). *Common core state standards for mathematics.* Washington, DC: Author. Retrieved from http://www.corestandards.org/assets/CCSSI_Math%20Standards.pdf

National Mathematics Advisory Panel. (2008). *Foundations for success: The final report of the National Mathematics Advisory Panel.* Washington, DC: U.S. Department of Education, Office of Planning, Evaluation and Policy Development.

National Research Council. (1996). *National science education standards.* Washington, DC: National Academy Press.

National Research Council. (2006). *Learning to think spatially: The incorporation of geographic information science across the K–12 curriculum.* Washington, DC: National Academy Press.

Newcombe, N. S., & Huttenlocher, J. (2000). *Making space: The development of spatial representation and reasoning.* Cambridge, MA: Massachusetts Institute of Technology.

Nunes, T., & Bryant, P. E. (1996). *Children doing mathematics.* Oxford, United Kingdom: Blackwell.

Olive, J. (2001). Children's number sequences: An explanation of Steffe's constructs and an extrapolation to rational numbers of arithmetic. *Mathematics Educator, 11*(1), 4–9.

Outhred, L., & Mitchelmore, M. (1992). Representation of area: A pictorial perspective. In W. Geeslin & K. Graham (Eds.), *Proceedings of the 16th Conference of the International Group for the Psychology in Mathematics Education* (Vol. 2, pp. 194–201). Durham, NH.

Outhred, L. N., & Mitchelmore, M. C. (2000). Young children's intuitive understanding of rectangular area measurement. *Journal for Research in Mathematics Education, 31*(2), 144–167. doi:10.2307/749749

Petrosino, A. J., Lehrer, R., & Schauble, L. (2003). Structuring error and experimental variation as distribution in the fourth grade. *Mathematical Thinking and Learning, 5*(2–3), 131–156. doi:10.1080/10986065.2003.9679997

Pettito, A. L. (1990). Development of numberline and measurement concepts. *Cognition and Instruction, 7*(1), 55–78. doi:10.1207/s1532690xci0701_3

Piaget, J. (1952). *The child's conception of number* (C. Gattegno & F. M. Hodgson, Trans.). London, United Kingdom: Routledge and Kegan Paul. (Original work published 1941)

Piaget, J., & Inhelder, B. (1967). *The child's conception of space* (F. J. Langdon & J. L. Lunzer, Trans.). New York, NY: Norton. (Original work published 1948)

Piaget, J., Inhelder, B., & Szeminska, A. (1960). *The child's conception of geometry* (E. A. Lunzer, Trans.). New York, NY: Norton.

Pirie, S., & Kieren, T. (1994). Growth in mathematical understanding: How can we characterize it and how can we represent it? *Educational Studies in Mathematics, 26*(2–3), 165–190. doi:10.1007/BF01273662

Praet, M., Titeca, D., Ceulemans, A., & Desoete, A. (2013). Language in the prediction of arithmetics in kindergarten and grade 1. *Learning and Individual Differences, 27*, 90–96. doi:10.1016/j.lindif.2013.07.003

Presmeg, N., & Barrett, J. (2003). Lesson study characterized as a multi-tiered teaching experiment. In N. A. Pateman, B. J. Dougherty, & J. T. Zillox (Eds.), *Proceedings of the Joint Meeting of the 25th Conference of the International Group for the Psychology of Mathematics Education (PME) and PME-NA* (Vol. 4, pp. 47–54). Honolulu, HI.

Purpura, D. J., & Ganley, C. M. (2014). Working memory and language: Skill-specific or domain-general relations to mathematics? *Journal of Experimental Child Psychology, 122*, 104–121. doi:10.1016/j.jecp.2013.12.009

Ravn, K. E., & Gelman, S. A. (1984). Rule usage in children's understanding of "big" and "little." *Child Development, 55*(6), 2141–2150. doi:10.2307/1129787

Reeve, B. B., & Fayers, P. (2005). Applying item response theory modelling for evaluating questionnaire items and scale properties. In P. M. Fayers & R. D. Hays (Eds.), *Assessing quality of life in clinical trials: Methods and practice* (2nd ed., pp. 55–73). New York, NY: Oxford University Press.

Resnick, L. B., & Ford, W. W. (1981). *The psychology of mathematics for instruction*. Mahwah, NJ: Erlbaum.

Rulence-Paques, P., & Mullet, E. (1998). Area judgment from width and height information: The case of the rectangle. *Journal of Experimental Child Psychology, 69*(1), 22–48. doi:10.1006/jecp.1998.2434

Russell, J. (1975). The interpretation of conservation instructions by five-year-old children. *Journal of Child Psychology and Psychiatry, 16*(3), 233–244. doi:10.1111/j.1469-7610.1975.tb01272.x

Ruthven, K., Laborde, C., Leach, J., & Tiberghien, A. (2009). Design tools in didactical research: Instrumenting the epistemological and cognitive aspects of the design of teaching sequences. *Educational Researcher, 38*(5), 329–342. doi:10.3102/0013189X09338513

Sandhofer, C. M., & Smith, L. B. (1999). Learning color words involves learning a system of mappings. *Developmental Psychology, 35*(3), 668–679. doi:10.1037/0012-1649.35.3.668

Sarama, J., & Clements, D. H. (2009a). "Concrete" computer manipulatives in mathematics education. *Child Development Perspectives, 3*(3), 145–150. doi:10.1111/j.1750-8606.2009.00095.x

Sarama, J., & Clements, D. H. (2009b). *Early childhood mathematics education research: Learning trajectories for young children.* New York, NY: Routledge.

Sarama, J., Clements, D. H., Barrett, J. E., Van Dine, D. W., & McDonel, J. S. (2011). Evaluation of a learning trajectory for length in the early years. *ZDM: The International Journal on Mathematics Education, 43*(5), 667–680. doi:10.1007/s11858-011-0326-5

Sarama, J., Clements, D. H., Van Dine, D. W., McDonel, J. S., Napora, L., & Barrett, J. E. (2011, April). *A hypothetical learning trajectory for volume in the early years.* Paper presented at the Annual Meeting of the American Educational Research Association, New Orleans, LA.

Saxe, G. B., Shaughnessy, M. M., Shannon, A., Langer-Osuna, J. M., Chinn, R., & Gearhart, M. (2007). Learning about fractions as points on a number line. In W. G. Martin & M. E. Strutchens (Eds.), *The learning of mathematics,* 2007 yearbook of the National Council of Teachers of Mathematics (pp. 221–238). Reston, VA: National Council of Teachers of Mathematics.

Schiff, W. (1983). Conservation of length redux: A perceptual-lingustic phenomenon. *Child Development, 54*(6), 1497–1506. doi:10.2307/1129812

Schumacher, R. F., & Fuchs, L. S. (2012). Does understanding relational terminology mediate effects of intervention on compare word problems? *Journal of Experimental Child Psychology, 111*(4), 607–628. doi:10.1016/j.jecp.2011.12.001

Sena, R., & Smith, L. B. (1990). New evidence on the development of the word *big*. *Child Development, 61*(4), 1034–1052. doi:10.1111/j.1467-8624.1990.tb02840.x

Shepard, L., Hammerness, K., Darling-Hammond, L., & Rust, F. (with Snowden, J. B., Gordon, E., Gutierrez, C., & Pacheco, A.). (2005). Assessment. In L. Darling-Hammond & J. Bransford (Eds.), *Preparing teachers for a changing world: What teachers should learn and be able to do* (pp. 275–326). San. Francisco, CA: Jossey-Bass.

Siegler, R. S., & Booth, J. L. (2004). Development of numerical estimation in young children. *Child Development, 75*(2), 428–444. doi:10.1111/j.1467-8624.2004.00684.x

Siegler, R. S., & Svetina, M. (2006). What leads children to adopt new strategies? A microgenetic/cross-sectional study of class inclusion. *Child Development, 77*(4), 997–1015. doi:10.1111/j.1467-8624.2006.00915.x

Silverman, I. W., York, K., & Zuidema, N. (1984). Area-matching strategies used by young children. *Journal of Experimental Child Psychology, 38*(3), 464–474. doi:10.1016/0022-0965(84)90089-4

Simon, M. A. (1995). Reconstructing mathematics pedagogy from a constructivist perspective. *Journal for Research in Mathematics Education, 26*(2), 114–145. doi:10.2307/749205

Simon, M. A. (2006). Key developmental understandings in mathematics: A direction for investigating and establishing learning goals. *Mathematical Thinking and Learning, 8*(4), 359–371. doi:10.1207/s15327833mtl0804_1

Simon, M. A. (2013). Issues in theorizing mathematics learning and teaching: A contrast between learning through activity and DNR research programs. *Journal of Mathematical Behavior, 32*(3), 281–294. doi:10.1016/j.jmathb.2013.03.001

Simon, M. A., & Tzur, R. (2004). Explicating the role of mathematical tasks in conceptual learning: An elaboration of the hypothetical learning trajectory. *Mathematical Thinking and Learning, 6*(2), 91–104. doi:10.1207/s15327833mtl0602_2

Smith, C. L., Wiser, M., Anderson, C. W., & Krajcik, J. S. (2006). Implications of research on children's learning for standards and assessment: A proposed learning progression for matter and the atomic-molecular theory. *Measurement: Interdisciplinary Research and Perspectives, 4*(1–2), 1–98. doi:10.1080/15366367.2006.9678570

Smith, J. P., III, Males, L. M., Dietiker, L. C., Lee, K., & Mosier, A. (2013). Curricular treatments of length measurement in the United States: Do they address known learning challenges? *Cognition and Instruction, 31*(4), 388–433. doi:10.1080/07370008.2013.828728

Smith, J. P., III, & Thompson, P. W. (2007). Quantitative reasoning and the development of algebraic reasoning. In J. J. Kaput, D. W. Carraher, & M. L. Blanton (Eds.), *Algebra in the early grades* (pp. 95–132). New York, NY: Erlbaum.

So, W. W.-M. (2013). Connecting mathematics in primary science inquiry projects. *International Journal of Science and Mathematics Education, 11*(2), 385–406. doi:10.1007/s10763-012-9342-3

Sophian, C. (2002). Learning about what fits: Preschool children's reasoning about effects of object size. *Journal for Research in Mathematics Education, 33*(4), 290–302. doi:10.2307/749742

Steffe, L. P. (1991). Operations that generate quantity. *Learning and Individual Differences, 3*(1), 61–82. doi:10.1016/1041-6080(91)90004-K

Steffe, L. P., & Cobb, P. (with von Glasersfeld, E.). (1988). *Construction of arithmetical meanings and strategies*. New York, NY: Springer-Verlag.

Steffe, L. P., & Olive, J. (2010). *Children's fractional knowledge*. New York, NY: Springer.

Steffe, L. P., & Thompson, P. W. (with von Glasersfeld, E.) (2000). Teaching experiment methodology: Underlying principles and essential elements. In A. E. Kelly & R. A. Lesh (Eds.), *Handbook of research design in mathematics and science education* (pp. 267–306). Mahwah, NJ: Erlbaum.

Stephan, M., Bowers, J., Cobb, P., & Gravemeijer, K. (2004). *Supporting students' development of measuring conceptions: Analyzing students' learning in social context*. Journal for Research in Mathematics Education monograph series (Vol. 12). Reston, VA: National Council of Teachers of Mathematics.

Stephan, M., & Clements, D. H. (2003). Linear and area measurement in prekindergarten to grade 2. In D. H. Clements (Ed.), *Learning and teaching measurement,* 2003 yearbook of the National Council of Teachers of Mathematics (pp. 3–16). Reston, VA: National Council of Teachers of Mathematics.

Sztajn, P., Confrey, J., Wilson, P. H., & Edgington, C. (2012). Learning trajectory based instruction: Toward a theory of teaching. *Educational Researcher, 41*(5), 147–156. doi:10.3102/0013189X 12442801

Tal, E. (2013). Old and new problems in philosophy of measurement. *Philosophy Compass, 8*(12), 1159–1173. doi:10.1111/phc3.12089

Tatsuoka, C., Clements, D. H., Sarama, J., Izsák, A., Orrill, C. H., de la Torre, J., . . . Tatsuoka, K. K. (2016). Developing workable attributes for psychometric models based on the Q-Matrix. In A. Izsák, J. T. Remillard & J. Templin (Eds.), *Psychometric methods in mathematics education: Opportunities, challenges, and interdisciplinary collaborations* (Monograph #15) (pp. 73-96). Reston, VA: National Council of Teachers of Mathematics.

Tatsuoka, K. K. (2009). *Cognitive assessment: An introduction to the rule space method*. New York, NY: Routledge.

Tharp, R. G., & Gallimore, R. (1988). *Rousing minds to life: Teaching, learning, and schooling in social contexts*. New York, NY: Cambridge University Press.

Thompson, A. G., & Thompson, P. W. (1996). Talking about rates conceptually, Part II: Mathematical knowledge for teaching. *Journal for Research in Mathematics Education, 27*(1), 2–24. doi:10.2307/749194

Thompson, P. W. (1992). Notations, conventions, and constraints: Contributions to effective uses of concrete materials in elementary mathematics. *Journal for Research in Mathematics Education, 23*(2), 123–147. doi:10.2307/749497

Thompson, P. W., & Thompson, A. G. (1994). Talking about rates conceptually, Part I: A teacher's struggle. *Journal for Research in Mathematics Education, 25*(3), 279–303. doi:10.2307/749339

Thorndike, E. L. (1922). *The psychology of arithmetic*. New York, NY: Macmillan.

Turner, E., Dominquez, H., Maldonado, L., & Empson, S. B. (2013). English learners' participation in mathematical discussion: Shifting positionings and dynamic identities. *Journal for Research in Mathematics Education, 44*(1), 199–234. doi:10.5951/jresematheduc.44.1.0199

van de Rijt, B. A. M., & Van Luit, J. E. H. (1999). Milestones in the development of infant numeracy. *Scandinavian Journal of Psychology, 40*(1), 65–71. doi:10.1111/1467-9450.00099

van den Heuvel-Panhuizen, M., & Buys, K. (Eds.). (2008). *Young children learn measurement and geometry: A learning-teaching trajectory with intermediate attainment targets for the lower grades in primary school.* Rotterdam, the Netherlands: Sense Publishers.

van Hiele, P. M. (1986). *Structure and insight: A theory of mathematics education.* Orlando, FL: Academic Press.

van Hiele-Geldof, D. (1984). The didactics of geometry in the lowest class of secondary school (M. Verdonck, Trans.). In D. Fuys, D. Geddes, & R. Tischler (Eds.), *English translation of selected writings of Dina van Hiele-Geldof and Pierre M. van Hiele* (pp. 1–214). Brooklyn, NY: Brooklyn College, School of Education. Retrieved from ERIC database. (ED289697)

Vergnaud, G. (1982). A classification of cognitive tasks and operations of thought involved in addition and subtraction problems. In T. P. Carpenter, J. M. Moser, and T. A. Romberg (Eds.), *Addition and subtraction: A cognitive perspective* (pp. 39–59). Hillsdale, NJ: Erlbaum.

Vukovic, R. K., & Lesaux, N. K. (2013). The language of mathematics: Investigating the ways language counts for children's mathematical development. *Journal of Experimental Child Psychology, 115(*2), 227–244. doi:10.1016/j.jecp.2013.02.002

Vygotsky, L. (1986). *Thought and language* (Rev. ed.; A. Kozulin, Ed.). Cambridge, MA: Massachusetts Institute of Technology. (Original work published 1934)

Vygotsky, L. S. (1978). *Mind in society: The development of higher psychological processes* (M. Cole, V. John-Steiner, S. Scribner, & E. Souberman, Eds.). London, England: Harvard University Press.

Wager, A. A., & Carpenter, T. P. (2012). Learning trajectories through a socio-cultural lens. In J. S. Carlson & J. R. Levin (Eds.), *Psychological perspectives on contemporary educational issues: Vol. 3. Instructional strategies for improving student learning: Focus on early mathematics and reading* (pp. 197–204). Charlotte, NC: Information Age Publishing.

Watson, J. M., Callingham, R. A., & Kelly, B. A. (2007). Students' appreciation of expectation and variation as a foundation for statistical understanding. *Mathematical Thinking and Learning, 9*(2), 83–130. doi:10.1080/10986060709336812

Wickstrom, M. H. (2014). *An examination of teachers' perceptions and implementation of learning trajectory based professional development* (Unpublished doctoral dissertation). Illinois State University, Normal, IL.

Wickstrom, M. H., Baek, J., Barrett, J. E., Cullen, C. J., & Tobias, J. M. (2012). Teacher's noticing of children's understanding of linear measurement. In L. R. Van Zoest, J.-J. Lo, & J. L. Kratky (Eds.), *Proceedings of the 34th Annual Meeting of the North American Chapter of the International Group for the Psychology of Mathematics Education* (pp. 488–494). Kalamazoo, MI.

Wilson, P. H. (2009). *Teachers' uses of a learning trajectory for equipartitioning* (Doctoral dissertation). Available from ProQuest Dissertations and Theses database. (UMI No. 3395288)

Wright, B. D., & Linacre, J. M. (1994). Reasonable mean-square fit values. *Rasch Measurement Transactions, 8*(3), 370. Retrieved from http://www.rasch.org/rmt/rmt83b.htm

Wright, B. D., & Stone, M. H. (1979). *Best test design: Rasch measurement.* Chicago, IL: Mesa Press.